Agents in the Long Game of AI

Agents in the Long Game of AI

Computational Cognitive Modeling for Trustworthy, Hybrid AI

Marjorie McShane, Sergei Nirenburg, and Jesse English

The MIT Press
Cambridge, Massachusetts
London, England

The MIT Press would like to thank the anonymous peer reviewers who provided comments on drafts of this book. The generous work of academic experts is essential for establishing the authority and quality of our publications. We acknowledge with gratitude the contributions of these otherwise uncredited readers.

This book was set in Stone Serif and Stone Sans by Westchester Publishing Services. Printed and bound in the United States of America.

Library of Congress Cataloging-in-Publication Data

Names: McShane, Marjorie Joan, 1967- author. | Nirenburg, Sergei, author. | English, Jesse, author.
Title: Agents in the long game of AI : computational cognitive modeling for trustworthy, hybrid AI / Marjorie McShane, Sergei Nirenburg, and Jesse English.
Description: Cambridge, Massachusetts : The MIT Press, [2024] | Includes bibliographical references and index.
Identifiers: LCCN 2023054788 (print) | LCCN 2023054789 (ebook) | ISBN 9780262549424 (paperback) | ISBN 9780262380348 (epub) | ISBN 9780262380355 (pdf)
Subjects: LCSH: Artificial intelligence. | Natural language processing (Computer science) | Computational linguistics.
Classification: LCC Q335 .M3925 2024 (print) | LCC Q335 (ebook) | DDC 006.3—dc23/eng/20240310
LC record available at https://lccn.loc.gov/2023054788
LC ebook record available at https://lccn.loc.gov/2023054789

10 9 8 7 6 5 4 3 2 1

Contents

Acknowledgments

Our sincerest thanks to Stephen Beale, Bruce Jarrell, and George Fantry, whose ideas continue to reverberate in our work; to Kenneth Goodman for spurring our thinking about explainability; to Sanjay Oruganti for his contributions involving large language models; to Nikhil Krishnaswamy for fruitful methodological discussions; to the Office of Naval Research for their generous support over many years;[1] and specifically to our program officers at the Office of Naval Research, Thomas McKenna and Michael K. Qin, for enabling our vision of artificial intelligence to become a reality.

1 Setting the Stage

For the past several decades, the field of artificial intelligence (AI) has been almost exclusively pursuing the hypothesis that machine learning is *the* path to developing systems with artificial intelligence. But what if machine learning is not the whole answer? What if the machine-learning-based technologies that have revolutionized modern life have limitations that make them unable to independently and reliably implement society's vision of an AI-enhanced future?

In the current climate, this is a disruptive notion. But it is more than just a notion—it is the rationale for developing the hybrid cognitive systems we call language-endowed intelligent agents (LEIAs). The LEIA program of research and development is a theoretically grounded, long-term, integrationist effort that has two main emphases: developing systems whose capabilities extend beyond what machine learning alone can offer and earning people's trust in those systems through explainability. LEIAs can explain their operation because they are configured using human-inspired computational cognitive modeling. Their explanations make clear the relative contributions of symbolic and data-driven methods, which is similar to a human doctor explaining a recommended procedure using both causal chains, such as how the procedure works, and population-level statistics, such as the percentage of patients for whom it is curative.

At the time of writing, large language models (LLMs) are animating AI professionals and the popular imagination alike. They are able to generate natural-sounding text and can often provide reasonable responses to prompts based on the word co-occurrence probabilities in massive datasets. This veneer of intelligence might give the impression that the language problem in AI—or even the problem of artificial intelligence overall—has been solved. But this isn't the case. Instead, LLMs are a tool that, if used judiciously, can contribute to the advancement of AI, including by speeding up LEIA development in ways detailed in the chapters to come.

Current LEIA development focuses on configuring agents that serve as collaborators with humans and, accordingly, share their goals. As collaborators, LEIAs will gain the trust of their human partners by (a) making competent and correct decisions and recommendations within their areas of expertise, (b) communicating their levels of confidence in those decisions and recommendations, (c) explaining those decisions and recommendations in maximally causal terms, and (d) learning over time through interaction. Agents that start out as strictly collaborative can, over time, be supplied with adversarial capabilities as well.

Distinguishing features of the LEIA program of R&D are as follows:

1. LEIAs are **social cognitive systems** capable of perception, reasoning, and action.

2. They can be implemented in **multimodal systems** that include multiple channels of perception, such as language and vision, and multiple types of action, such as language generation and simulated and robotic action.

3. LEIAs orient around **meaning**, which is defined with respect to an unambiguous, property-rich, script-enhanced ontology (world model).

4. When LEIAs interpret any kind of perceptual input, they represent its meaning using a standard kind of **symbolic meaning representation**, which is stored to memory and then used in reasoning about action. This means that a LEIA's memory contains digested data that has been analyzed, organized, categorized, grounded, and interconnected so that it becomes **knowledge**.

5. LEIAs use **large knowledge bases** that include both **commonsense and expert knowledge**. This means that, from the outset, we are addressing the theoretical and practical challenges of **scaling up**.

6. The combined emphases on meaning and large knowledge bases make LEIAs **content-centric** intelligent agents.

7. A LEIA's ontology and most aspects of its reasoning are language-independent and, therefore, applicable **crosslinguistically**.

8. The work of developing LEIAs involves **acquiring knowledge in conjunction with developing the reasoners that will use it**.

9. LEIAs are particularly strong in **meaning-oriented language under-standing and generation**, which is key to **communicating, learning,** and **explaining**. The moniker "LEIA" (language-endowed intelligent agent) reflects the importance of language processing for LEIAs with no implication that language is their exclusive concern.

10. A cognitive model of explanation enables LEIAs to generate the kinds of **intuitive, understandable explanations** that humans want and need.

11. In order to exploit the fruits of data-driven AI (machine learning and data analytics), LEIAs are **hybrid** systems. When they explain reasoning that incorporates data-driven evidence, they identify the role of that evidence.

12. The LEIA program of work incorporates both **basic science** and **technology**. As science, it involves building human-inspired theories and models, which includes explaining complex aspects of human cognition. As technology, it involves building systems that are useful in the near term as they evolve over time. The goal is to work at the sweet spot between science and technology in order to solve the long-term problems of AI in a practical way.

13. LEIA development combines **domain-specific** and **domain-independent capabilities**. Knowledge and functionalities are as domain-independent as possible but as domain-specific as necessary.

14. LEIAs rely on **knowledge representation** methods that integrate static and procedural knowledge in a clear, traceable, and extensible manner. Advancements in knowledge representation strategies separate LEIA development from the expert systems of early AI.

15. The **control programs** that operate over the knowledge are **maximally compact**, thus avoiding the pitfalls of mixing knowledge into code, which results in impenetrable code bases. Agents become more sophisticated primarily by expanding their knowledge bases, not their code bases.

16. **Knowledge acquisition methodologies** combine manual, semiautomatic, and fully automatic modes as a practical path toward overcoming what some have called *the knowledge bottleneck*. This, however, is a misnomer: there is no bottleneck—there is simply work that needs to be done.

17. LEIAs are being designed to serve as **collaborators** with, rather than replacements for, people. The division of labor has LEIAs take over the

tasks they are best suited for, leaving people to do what they can do easily and what they must be responsible for. The latter **mitigates ethical concerns.**

18. LEIAs target **applications that cannot be adequately served by machine learning.** Applications for which machine learning is suitable have been at the center of people's attention and have profoundly affected daily life. However, many kinds of applications are not being attempted because machine learning does not suffice. It is these kinds of applications that LEIA development seeks to serve.

19. All of the reported algorithms are computer-tractable, and most of them have been **implemented** as prototypes, in keeping with our status as an academic research lab. Implementation is essential to testing ideas and algorithms because the demands of system building highlight algorithmic deficiencies. The solutions vetted in our prototypes can then be expanded on, in a theme-and-variations spirit, by non-research technologists.

20. Any cognitive system that is developed according to the theoretical and methodological principles described in this book is, in our parlance, a LEIA. This book describes **why and how to develop LEIAs**, using our research group's implementation as an illustration.

This book is the third in a sequence addressing explainable, meaning-oriented AI. *Ontological Semantics* (Nirenburg & Raskin, 2004) lays the foundation for ontologically grounded language processing, and its successor, *Linguistics for the Age of AI* (McShane & Nirenburg, 2021), presents advances within the context of agent systems. The current book describes our variety of agents more holistically, as hybrid, multifunctional cognitive systems whose computational methods enable new kinds of AI capabilities—including the ability of agent systems to provide the kinds of explanations that will earn humans' trust.

*   *   *

A terminological note: in passages about LEIAs, the terms *agent* and *LEIA* are interchangeable, as are the terms *model* and *microtheory*.

An online appendix is available at https://faculty.rpi.edu/marjorie-mcshane.

2 Content-Centric Cognitive Modeling

The language-endowed intelligent agents (LEIAs) described in this book belong to a family of cognitive systems that aim to emulate the human abilities of perception, reasoning, and action. But LEIA research differs from most cognitive systems research with respect to the following parameters:

1. LEIAs are designed to use large-scale stored knowledge and self-generated situational knowledge to support their reasoning about perception, decision-making, and action.

2. LEIAs are designed to engage in lifelong, humanlike learning, which involves acquiring knowledge by reading and through show-and-tell interactions with humans.

3. LEIAs are prepared to explain their knowledge, reasoning, and operation, with the goal of engendering the trust of their human partners.

It is our emphasis on knowledge content that explains the coinage *content-centric cognitive modeling*.[1]

The basic principles of LEIA modeling were listed in chapter 1. This chapter introduces the cognitive architecture underlying LEIAs, outlines their symbolic-empirical hybridization, and describes choices and methodologies underpinning this program of research and development. Comparisons with other approaches are presented in the final section.

2.1 The OntoAgent Cognitive Architecture

Our research group develops LEIAs within a cognitive architecture called OntoAgent. Figure 2.1 presents a high-level sketch. For clarity of presentation, it does not show all processing flows and only hints at the central role played by the agents' knowledge bases throughout their operation.

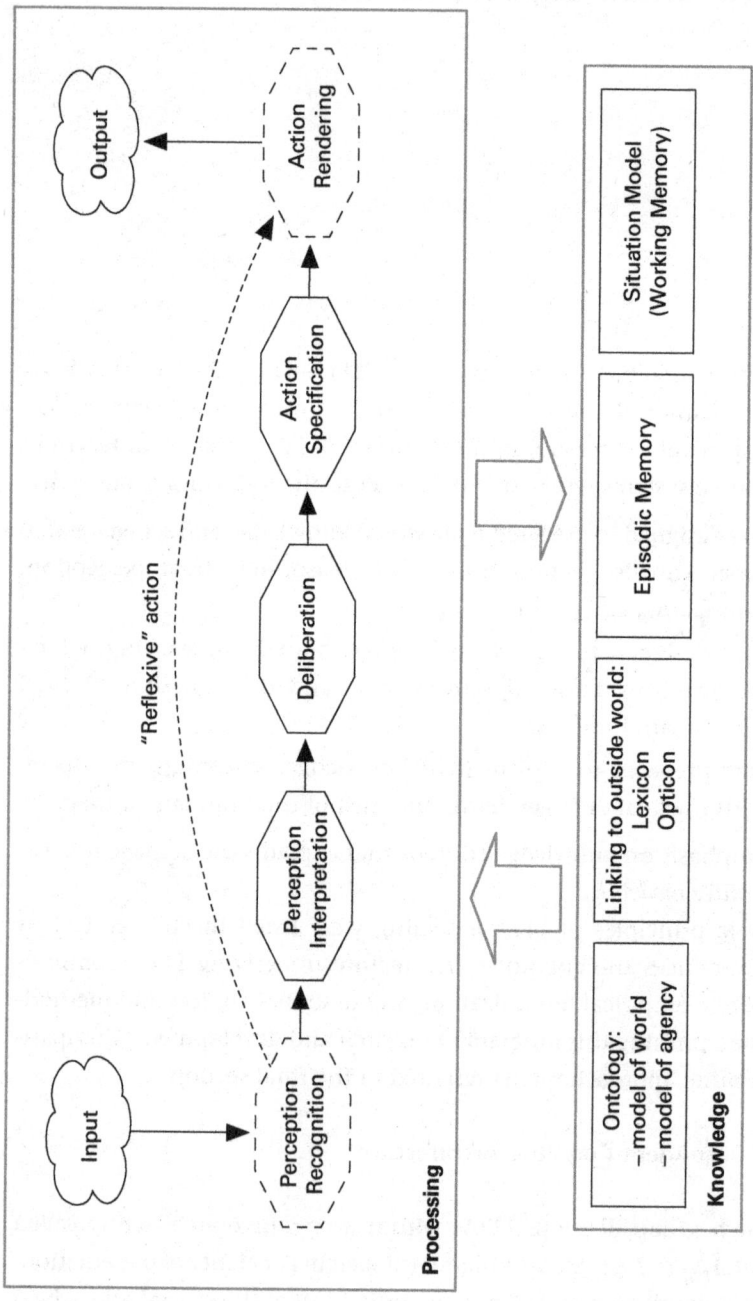

Figure 2.1

A high-level sketch of OntoAgent, the cognitive architecture supporting the development of LEIAs. In the Perception Recognition and Action Rendering modules, data-driven methods play a significant role.

OntoAgent consists of:

1. a set of processing modules—the five octagons in figure 2.1, which are described in sections 2.1.1–2.1.5;

2. a knowledge substrate whose main components are: the ontology; knowledge bases that map elements of the outside world to ontological concepts, such as the lexicon and the opticon; episodic memory; and the situation model, also called working memory; and

3. infrastructure components that support both system functioning (e.g., the code base) and system development (e.g., DEKADE: the Development, Evaluation, Knowledge Acquisition, and Demonstration Environment; cf. section 2.5.2).

The knowledge bases and processing modules are modeled using the ontological metalanguage, which enables metacognition by LEIAs, including the ability to explain their own functioning. Agents process inputs by instantiating the ontological script called AGENT-FUNCTIONING-FLOW, which guides them through the five stages of processing using a sequence of events that they understand, remember, and can explain. This script is described in section 3.2.4, where it is used to illustrate the content and utility of ontological scripts.

In addition to following the five-stage, reasoning-intensive processing flow, agents can also undertake "reflexive" action, shown by the dotted arrow in figure 2.1, which simulates physiological reflexes as well as reasoning by analogy. Reasoning-intensive and reflexive processing have been referred to in the psychological literature as *slow thinking* and *fast thinking*, respectively (see, e.g., Kahneman, 2011).[2] Choosing between slow and fast thinking is the responsibility of attention management, which scopes over and informs the agent's basic functioning flow.

The OntoAgent architecture maintains interoperability across its Perception, Deliberation, and Action modules by (a) using a common, ontologically grounded meaning representation language across knowledge bases and internal processors, and (b) making the results of processing available to all modules of the architecture.

To-date, proof-of-concept systems developed on the basis of OntoAgent have demonstrated many functionalities, including language understanding and generation, interpreting the results of visual perception, simulated interoception (i.e., the experiencing of bodily signals), many types of reasoning and decision-making, learning, simulated action, robotic motor action, and simulated human physiology.

We will now work through the five stages of the processing module of OntoAgent.

2.1.1 Perception Recognition

When a LEIA attends to a new input, it begins with Perception Recognition, which includes two subtasks: detecting the type of input (such as speech, text, or vision) and preparing it for semantic analysis. Perception Recognition incorporates data-driven system modules. The particulars of hybridization, as well as outstanding challenges, differ across perception modalities.

Speech input Speech must be converted into text before it can be semantically analyzed. Speech-to-text conversion is a task for which machine learning is well suited, and good speech-to-text systems are available off the shelf. However, transcription errors are common, particularly if there are overlapping speakers, background noise, fragmentary utterances, or language that contains proper nouns or specialist terminology—a list unsurprising to those familiar with auto-generated closed captioning. In addition, speech-to-text systems do not address prosodic features to the extent needed for full semantic analysis.[3] For example, the sequence of words *Tony drove to the store* requires different interpretations depending on intonation and emphasis:

(2.1) a. Tony drove to the store.   [An assertion.]
       b. *Tony* drove to the store?   [It was Tony, not someone else?]
       c. Tony drove to *the store?*   [He didn't go somewhere else?]
       d. Tony *drove* to the store?   [He didn't get there some other way?]
       e. Tony drove to the store??!!   [I am shocked or outraged.]

During Perception Recognition, speech-to-text systems need to decorate text segments with linguistically informed prosodic features that can then be interpreted by LEIAs as part of Perception Interpretation. Prosodic feature recognition might well be amenable to machine-learning methods but remains, as yet, understudied (mainstream AI does not pursue a level of semantic or pragmatic analysis that requires it).

Text input Although text might seem like straightforward input to semantic analysis, complexity lurks just beneath the surface. All meaningful typographic conventions need to be semantically or functionally interpreted: italics, boldface, underlining, headings, lists, tables, speaker turns in plays, figure captions, and so on. In mainstream natural language processing, recognizing such features is traditionally handled by a preprocessor. In fact, automatically tagging such features was one of the goals of the Semantic Web push in the early aughts.[4] In texts that are not decorated by hypertext tags, all of these features need to be identified and made explicit as input to semantic analysis.

Vision In simulated worlds, visual input can be recognized using symbolic or data-driven methods. For example, in a past driving application, LEIAs serving as autonomous-vehicle operators recognized images based on their symbolic encoding in the simulation environment.[5] When objects are recognized this way, Perception Recognition and Perception Interpretation are folded together, since the symbolic encoding *is* the meaningful interpretation of the object. By contrast, in robotics applications, images and scenes are recognized using visual recognition systems that generate data that must then be converted by LEIAs into ontologically grounded meaning representations. This process is mediated by an opticon, which maps the output of data-driven Perception Recognition systems to ontological concepts in a way analogous to what a lexicon does for language interpretation (cf. section 3.4).[6]

Interoception Interoception is the experiencing of one's bodily signals— physical, psychological, and emotional. Since interoception affects decision-making, it must be modeled in embodied LEIAs. In a past project, LEIAs playing the role of virtual patients interpreted the physiological signals produced by the simulation system differently depending on their mental and emotional states.[7] Some patients, like those under great stress or suffering from hypochondria, interpreted objectively mild symptoms as intense, which could influence how the clinical scenario played out.

The interoception processor converts objective physiological signals into an agent's experiences based on relevant feature values in the agent's model of self. It is noteworthy that the interoception model need not be very complex to generate useful differentiation across agents and situations. It will be interesting to see how advances in neuroscience might inform this model, as by providing evidence about how much and why people differ in their experiences of objectively similar stimuli.

Other channels of perception Although our practical work on implementing LEIAs has not yet addressed all channels of perception, such as non-speech audio and haptics, the ones we have implemented provide a blueprint for the others. For example, if a robot needs to halt when it hears any loud noise, it needs to recognize certain audio inputs as instances of the ontological concept LOUD-NOISE and then launch either the decision or the reflex (depending on how this reaction is implemented) to halt. Similarly, if somebody says, in response to the sound of a goldfinch singing, "Listen to that," then *that* refers to interpreted audio input that must map to the concept GOLDFINCH-SINGING. Anything else that can be detected using sensors can be treated similarly. For example, if medical monitors are to serve as input to an automatic system for assessing a patient's status and issuing alerts, then the output of each must be translated into an associated ontological representation to be used in the agent's reasoning.

2.1.2 Perception Interpretation

Perception Interpretation is computing meaning—which raises the question, "What is meaning?" Although defining meaning has challenged scholars for centuries, one must commit to some definition in order to do the practical work of computational cognitive modeling. In the theory of Ontological Semantics that underlies LEIAs (Nirenburg & Raskin, 2004), the meaning of entities the agent perceives is modeled using knowledge structures encoded using an unambiguous, ontologically grounded metalanguage.[8] Agents reason, learn, and remember in terms of these knowledge structures.

Benefits of using ontologically grounded meaning representations include:

- They are unambiguous and fully specified.
- They are comprised of ontological concepts, each of which is described in the ontology using properties. So, the agent reasons using not only what it perceives but also everything it knows about the perceived kinds of objects and events. For example, to understand why a person would say, "Look, that car has no bumpers," one has to know that cars usually have bumpers.
- They are formally the same, apart from metadata, no matter which channel of perception gave rise to them.
- They are language independent, which means that the majority of agent knowledge and agent operation can be applied in any language setting.

Automatically translating between real-world data and interpreted knowledge is difficult, but it is exactly what is needed to create the next generation of agents that will be substantially more humanlike than today's AI because they will achieve a functional approximation of human understanding.[9]

Different types of meaning representations, which are variations on a theme, participate in different aspects of agent functioning.

- **Text meaning representations** (TMRs) record the meaning of speech and text inputs. They are computed during natural language understanding.

- **Vision meaning representations** (VMRs) record the interpretation of visual stimuli. They are computed during vision interpretation.

- **Interoception meaning representations** (IMRs) record the agent's interpretations of its simulated or robotic bodily signals. They are computed during interoception interpretation.

- **Generation meaning representations** (GMRs) record content specifications for language generation. They are computed as part of reasoning about action.

- **Action meaning representations** (AMRs) record content specifications for physical actions. They, too, are computed as part of reasoning about action.

- **Mental meaning representations** (MMRs) record results of reasoning that the agent remembers but does not act on externally.

- And so on, for any other types of inputs, outputs, or internal reasoning activities in the agent environment.

All meaning representations—referred to generically as XMRs—share the same ontological metalanguage and a generic set of properties.[10] So, no matter how an agent learns about something, or no matter why it generates a thought about something, the basic shape and content of the meaning representation is the same. For example, the following XMR expresses the idea that a cat named Pogo is looking at a small, friendly dog. The indices indicate that these are instances of ontological concepts.

VOLUNTARY-VISUAL-EVENT-1
    AGENT   CAT-1
    THEME   DOG-1
    TIME   *find-anchor-time*
CAT-1
    AGENT-OF   VOLUNTARY-VISUAL-EVENT-1
    HAS-NAME   "Pogo"

DOG-1
    THEME-OF   VOLUNTARY-VISUAL-EVENT-1
    FRIENDLINESS   .8
    SIZE   .2

This meaning representation is read as follows.

- The first frame is headed by an instance of the concept VOLUNTARY-VISUAL-EVENT. Concepts are distinguished from words of English by the use of small caps. This is not vacuous upper-case semantics[11] because ontological concepts are defined using properties that support reasoning about language and the world.

- VOLUNTARY-VISUAL-EVENT-1 has three contextually relevant property values: its AGENT (the one looking) is an instance of CAT; its THEME (the one being looked at) is an instance of DOG; and the TIME of the event is the time of speech. The filler for the latter, *find-anchor-time*, is a call to a procedural semantic routine that can be run if temporal grounding is needed.

- The next frame, headed by CAT-1, shows not only the inverse (AGENT-OF) relation to VOLUNTARY-VISUAL-EVENT-1 but also that the name of this cat is Pogo.

- The next frame, headed by DOG-1, shows not only the inverse (THEME-OF) relation to VOLUNTARY-VISUAL-EVENT-1 but also two features of the dog: it is small—.2 on the abstract {0–1} scale of SIZE; and it is friendly—.8 on the abstract scale of FRIENDLINESS. By convention, unmodified high or low property values are .2 away from the high or low extreme of the abstract {0,1} scale, which leaves room for interpreting expressions like *very friendly* (FRIENDLINESS .9) *and extremely friendly* (FRIENDLINESS 1).

Note that the concept FRIENDLINESS reflects a more generic notion than the English word *friendly* and supports the semantic interpretation of a large class of modifiers; for example, the meaning of *hostile* is represented as FRIENDLINESS with the value 0. This approach to recording the meaning of scalar attributes avoids an unnecessary proliferation of ontological concepts, in the spirit of Hayes (1979).

The above meaning representation is an abstraction: As we said, it expresses *the idea that a cat named Pogo is looking at a small, friendly dog.* If it were generated from an actual act of perception, it would have contained metadata with traces of agent reasoning. For example, if it were generated from the language input *A small friendly dog is being watched by Pogo the cat*, then it would be a *text* meaning representation (TMR) that indicated, for each frame, the word number that the frame is analyzing (*word-num*), the word itself (*textstring*), the lexical sense that was used to generate the analysis (*lex-sense*), and other words and lexical senses used in the frame—typically as modifiers (*uses-word, uses-lex-sense*).

> All knowledge structures in this book are presented in a reader-friendly format, with details being included or excluded based on the point they are being used to illustrate. Select aspects of meaning representations are explained inline, in plain English, preceded by a semi-colon. Examples of the actual knowledge structures used by LEIAs are presented in the online supplement. As regards meaning representations in particular, they are generally presented with little or no metadata in order to focus on the semantic content that the agent uses for reasoning.

A$_0$ small$_1$ friendly$_2$ dog$_3$ is$_4$ being$_5$ watched$_6$ by$_7$ Pogo$_8$ the$_9$ cat$_{10}$

VOLUNTARY-VISUAL-EVENT-1

AGENT	CAT-1	
THEME	DOG-1	
TIME	*find-anchor-time*	
word-num	6	
textstring	watched	
lex-sense	watch-v1	
uses-word	4, 5, 7	; folded into the passive transformation

CAT-1

AGENT-OF	VOLUNTARY-VISUAL-EVENT-1	
HAS-NAME	"Pogo"	
word-num	9	
textstring	the	
lex-sense	the-det5	; the construction "NP$_{NAME}$ the NP$_{ANIMAL/SOCIAL-ROLE}$"
uses-word	8, 10	; Pogo, cat
uses-lex-sense	cat-n1	
uses-lex-sense	personal-name-n1	; for "Pogo"

DOG-1

THEME-OF	VOLUNTARY-VISUAL-EVENT-1	
FRIENDLINESS	.8	
SIZE	.2	
COREF	block-coref	; because of 'a'
DISCOURSE-RELATION	TOPIC	; since *dog* is the subject of a passive clause
word-num	3	

textstring	dog
lex-sense	dog-n1
uses-word	0, 1, 2
uses-lex-sense	a-det1
uses-lex-sense	small-adj1
uses-lex-sense	friendly-adj1

By contrast, if a robotic agent learned this information by viewing the scene, then it would generate a VMR that would be decorated with information about which elements of the opticon were used to recognize all elements of this scene, including the specific instance of the cat that the agent knows to be named Pogo.

Meaning representations are integrated into the agent's long-term memory as an aspect of learning (see chapter 7).[12] If the information is generic, it enhances the ontology; otherwise, it becomes part of episodic memory. Integrating new information into the existing knowledge bases requires a battery of decisions, such as: Does the information pertain to an already-known object or event, or should a new anchor for it be established in memory? What should be done with new information that conflicts with existing knowledge? How and how often should repeated activities be transformed into generalizations about them, such as the fact that a particular person drinks coffee every day? This decision-making is carried out as part of memory management.

Automatically generating meaning representations is challenging, no matter their source. The associated processors—the natural language understanding system, the vision analyzer, the interoception engine, and so on—rely on knowledge and skills provided by a combination of static knowledge resources, symbolic reasoners, and data-driven tools.

2.1.3 Deliberation

The middle module of the OntoAgent architecture, which we call *Deliberation,* covers the agent's reasoning in service of planning, decision-making, learning, advising, memory management, and so on. We do not call this module *reasoning*—despite the traditional *perception-reasoning-action* terminology of cognitive architectures—because LEIAs reason during all stages of their operation except those that are outsourced to data-driven tools. Even then, LEIAs have to reason about the *results* of the outsourced processing by assigning them a confidence level and determining how to incorporate them into overarching symbolic reasoning. As concerns the complexity of different kinds of reasoning, reasoning about how to interpret perceptual inputs (carried out during Perception Interpretation) is every bit as difficult as reasoning about what to do next (carried out during Deliberation).

Calling this module *Deliberation* rather than *Reasoning* aims to preclude the reader from automatically making inferences about the nature of this module in the LEIA's architecture. However, it would be unnatural to avoid the term *reasoning* altogether. So, we trust that readers will take to heart the pervasiveness of reasoning throughout LEIA operation and will readily understand the context-specific implications of each use of the term *reasoning*.

LEIA reasoning emulates that of people in important ways. First, it takes into account beliefs, biases, intuitions, decision-making heuristics, and other less-than-perfectly-rational facets of human functioning.[13] Second, it orients around actionability. This means that LEIAs are designed to detect when they have understood a situation well enough to act. In high-risk situations, the bar for actionability will be very high, whereas in low-risk situations, the agent can chance a misstep, which can subsequently be corrected through interaction with its human partner. Finally, LEIA reasoning deemphasizes reasoning from first principles, which was a main focus of early AI. As a consequence, the efficiency of search is not a central concern of LEIAs. Instead, they pursue their goals using stored plans. Although people are certainly capable of search, they seem to use it in rather limited contexts, outside of game playing. Accordingly, content-centric modeling emphasizes the use of remembered, habitual solutions for heuristic decision-making. When no known plan leads to a goal, LEIAs, as social agents, prefer to ask for help rather than explore independently. This might be viewed as a facet of the cognitive miser theory (Stanovich, 2009), which posits that the *principle of least effort* applies to cognitive effort no less than to other human endeavors. Preferring communication to exploration in order to economize effort is only viable for language-endowed social agents. If collaborating with a human is, for some reason, not possible at a given moment, then the agent may opt to engage in reasoning from first principles, as by modifying a known plan using partial matching and reasoning by analogy. For example, remembering how lenses can be used to start a fire with sunlight allowed the modern Robinson Crusoe, Alvaro Cerezo, to use a plastic bag filled with sea water for this purpose.[14] He did not need to build a plan of action from scratch; he just amended a known plan by fulfilling a prerequisite—the availability of a lens—in a novel way. He could do this because of his understanding of the nature of lenses. In sum, content-centric modeling strives to replace search by retrieval from structured knowledge. It underscores the centrality of heuristics and downplays the dynamic generation of search spaces and search as such.

The kinds of reasoning undertaken in the Deliberation module are best explained using examples, which are amply provided in chapters 6–8. For comparison, reasoning in data-driven AI is discussed in section 2.7.1.

2.1.4   Action Specification
The output of the Deliberation module is the agent's decision about what to do, but not yet how to do it. How to do it is determined in the Action Specification module. For example, when asked a question, the agent must decide how many and which details to provide in the answer, which mode of communication to use (language and/or body language), which stylistic features to express (politeness, formality, enthusiasm), and which actual sentences and/or gestures will best serve in the given context. All of these aspects of Action Specification will be illustrated in upcoming chapters using examples of dialog interactions. The output of Action Specification is a meaning representation that needs to be rendered as an actual action using the appropriate effector, such as the text generator, the speech synthesizer, or robotic movement effectors.

2.1.5   Action Rendering
Action rendering is most naturally implemented using data-driven methods developed in fields such as robotics, video simulation, and text-to-speech generation. As such, action rendering lies outside of a LEIA's cognitively modeled core. However, the results of cognitive processing must be translated into the formats expected by data-driven action generation modules, just as the results of data-driven processing must be cognitively interpreted at the Perception Interpretation stage. For natural language generation, Action Rendering is divided into multiple stages that use different data-driven tools with different rationales and expectations of their performance (cf. section 4.3); this is a good example of how symbolic and data-driven approaches are being strategically integrated in LEIAs.

2.2 Hybridization

As we have seen, data-driven tools play a significant role at the peripheries of the LEIA's architecture, in the modules called Perception Recognition and Action Rendering. Data-driven tools also play a supporting role elsewhere. Their contribution is not only expedient but also aligns with principles of cognitive modeling related to (a) the distinction between skills and deliberative processes, and (b) the role of habits in human functioning.

As mentioned earlier, human reasoning can be viewed as a combination of *thinking fast* and *thinking slow*—also referred to as System 1 and System 2 (Kahneman, 2011). Thinking fast covers reflexive, habitual skills, whereas thinking slow covers deliberative processes. In terms of AI, as the coarsest generalization, data-driven tools implement thinking fast, whereas symbolic methods implement thinking slow. However, the real picture is considerably more nuanced.

Skills are behaviors that do not involve conscious thinking or decision-making, such as walking, picking up a cup, or recognizing a tree. Systems grounded in machine learning (ML)—particularly deep learning—have proven useful for emulating certain human skills in robots and simulated systems. What makes ML particularly suitable for implementing skills is that, in the general case, they don't need to be explained as long as they are performed reliably. However, this does not mean that ML-generated skills are entirely comparable to human skills. For example, people can deconstruct their skills in order to improve their performance, teach others, or adapt to nontraditional circumstances. Tennis pros practicing their serve, truck drivers going downhill in icy conditions, and biologists teaching students to distinguish between alligators and crocodiles are invoking conscious thinking for activities that they would typically carry out automatically, and this thinking involves concepts like follow-through, breaking distance, and snout shape. To the extent that ML-based systems can emulate skills accurately, they are useful. However, they still cannot teach people how to acquire or improve skills, nor can they understand or correct mistakes in the streamlined, one-shot-learning manner that humans can.

Habits result from the repetition of actions that initially involve thinking slow but, over time, turn into thinking-fast shortcuts. For example, someone might always prepare coffee using the same ordered sequence of events as long as there is nothing unusual about the situation. Modeling habits symbolically can be approached in two ways. On the one hand, the agent can actually experience a sequence of repeated, deliberate actions and, at some point, decide to encapsulate them into a habit. This is done using a memory consolidation function called CREATE-HABIT that is part of the Deliberation module. Specifically, when the CREATE-HABIT script is instantiated, it generates a new script that contains the given sequence of actions along with the prerequisite that the overall situation is normal. The new script retains traces of the original, slow-thinking reasoning, in case that should be needed for explanation or replanning. This script generation is similar to script learning by LEIAs in dialog applications.[15]

The other way an agent can be supplied with a habit is directly through knowledge engineering. For example, in dialog applications, rather than have agents go through semantic analysis, reasoning, and planning in order to respond to the greeting "Hi," the model can assert that hearing "Hi" reflexively results in responding "Hi" (see the dotted arrow in figure 2.1). Modeling such responses as reflexes not only emulates human behavior but also saves valuable time in both knowledge engineering and system operation without negatively impacting explainability. If explaining a *Hi—Hi* dialog pair *was* deemed important, then the full reasoning process could be modeled.

Deliberative (slow-thinking) reasoning, which offers explanatory power, is the forte of computational cognitive modeling. While it is true that some processes that, for humans, would be deliberative have been automated using ML—such as playing chess and making medical diagnoses—the operation of these systems has nothing in common with human reasoning.[16] Such systems are best thought of as oracles whose results can be useful in two cases: in domains where explanation is not needed, such as chess practice without coaching, and when the oracle's results will serve as heuristic evidence for more holistic decision-making, such as diagnosing a disease as part of the comprehensive medical care of a patient. In the latter case, the agent's confidence in, and explanation of, its holistically approached decision is affected by the opaqueness of the ML-based operations.

Modeling LEIAs as hybrid, rather than purely symbolic, systems makes sense on multiple counts.

1. Some functionalities, such as robotic movement, will likely never need to be explained, so there is no reason *not* to implement them using ML.

2. Some functionalities, such as medical image diagnostics, are inherently dependent on pattern recognition and are natively implemented using ML.

3. Some functionalities straddle the line between explainable and not explainable. For example, an agent's data-driven vision system might be able to reliably differentiate between cats and dogs but not be able to explain its choice in terms of the differentiating features people could easily point out, such as snout length, ear shape, and tail flexibility. Over time, it would be useful to develop hybrid vision recognition systems that addressed both the overall object or scene and its parts, with those decompositions mirroring the associated descriptions in the ontology.

4. Some data-driven functionalities can provide useful input for agent reasoning without having the responsibility of being an end product. For example, large language models (LLMs) are useful for preparing learning material for LEIAs.

5. Some data-driven functionalities help a LEIA to select from among multiple valid options; so, even if the agent's choice is not perfect, what is produced will not be far off. An example of this is using an LLM-based system to select the most contextually appropriate sentence from a set of semantically correct candidates produced by the LEIA (cf. section 4.3.5).

6. Finally, although LEIA modeling has a significant research orientation, it is a proper part of the practical discipline of AI. As such, implementing useful systems in finite time is an objective, and this objective is served by well-selected hybridization.

2.3 The Overall Methodology of LEIA Research and Development

Methodological choices for computational cognitive modeling can be traced using a *theory-model-system* trichotomy. As a first approximation, theories in cognitive science are abstract and formal statements about how human cognition works; models account for real data in computable ways and are influenced as much by practical considerations as by theoretical insights; and systems implement models within the real-world constraints of existing technologies. This trichotomy is at the heart of LEIA development and warrants further discussion.[17]

Theories. Theories aim to explain and reflect reality as it is, albeit with great latitude for underspecification. They are not bound by practical concerns such as computability or the attainability of prerequisites for processes they describe or explain. We share the position, formulated in Winther (2016, section 4.1.1) and attributed to Nancy Cartwright, that "laws of nature are rarely true and epistemically weak. Theory as a collection of laws cannot, therefore, support the many kinds of inferences and explanations that we have come to expect it to license." This position ascends to Cartwright's (1983) view that "to explain a phenomenon is to find a model that fits it into the basic framework of the theory and that thus allows us to derive analogues for the messy and complicated phenomenological laws which are true of it" (p. 152). Theories guide developers' thinking in developing models and interpreting their nature, output, and expectations. One of the theories underpinning LEIA development is the theory of Ontological Semantics, which informs all aspects of language processing by LEIAs (Nirenburg & Raskin, 2004).

Models. Computational cognitive models describe specific phenomena and methods for LEIAs to treat them.[18] Models must be computable, relying exclusively on types of input (e.g., property values) that can actually be computed using technologies and knowledge repositories available at the time of model construction. If some feature that plays a key role in a theory cannot be computed, then it either must be replaced by a computable proxy, if such exists, or it must be excluded from the model.[19] In other words, models, unlike theories, must include concrete decision algorithms and computable heuristics.

Models must account for the widest possible swath of data but at the same time embrace carefully selected simplifications to remain useful in practice. As past work in human-inspired machine reasoning has shown, it is counterproductive to construct decision functions with large sets of parameters; simple heuristics are often preferable.[20]

Models must reflect the distinction between competence and performance. For example, although native speakers of a language can distinguish between grammatical and ungrammatical sentences (they have language *competence*), far from every utterance people produce is grammatical (language *performance* is far from perfect). So computational cognitive models of language must mimic human success in natural, imperfect communication. More generally, LEIAs need to mimic human behavior in novel situations that are not completely understood. It follows from the above that they are liable to make less-than-optimal decisions at each of their processing stages. The key is for the underlying cognitive models to be able to explain why this happens and, over time, evolve to be increasingly fail-safe.

The notions of *cognitive load* and *actionability* are key to the computational cognitive modeling of LEIAs. Cognitive load describes how much effort people have to expend to carry out a mental task. As a first approximation, a low cognitive load for people should translate into a simpler processing challenge for machines and, accordingly, a higher confidence in the outcome. Of course, this is an idealization since certain analysis tasks that are simple for people—such as detecting sarcasm—can be difficult for machines. However, the basic insight remains valid: LEIAs must be aware of the cognitive load of their processing and have both a standing goal and a corresponding set of plans for keeping their cognitive load in check.

Actionability, for its part, captures the idea that people can often get by with an imperfect and incomplete understanding of both language and situations. LEIAs must imitate the human ability to judge whether they understand a particular input sufficiently to adequately respond to it or whether some repair activity is in order, such as initiating a clarification dialog with a human team member.

The agent's assessment of cognitive load, confidence, and actionability plays out differently in different contexts. For example, if an agent detects that its human collaborators are engaging in off-topic chitchat among themselves, it can consider its possibly poor analyses of those utterances *actionable* with the action being *ignore the utterances*. By contrast, if a LEIA is supporting a surgeon, anything less than full confidence in the interpretation of an order will necessarily lead to a clarification subdialog to avoid a potentially catastrophic error.

Finally, models must operationalize the factors identified as most important by the theory. Cognitive load and actionability provide useful illustrations of this requirement. The cognitive load of interpreting a given input can be estimated using a function that considers the number and complexity of each contributing task. Consider one example from each end of the complexity spectrum in language understanding. The sentence *Jessica ate an apple* will result in a low-complexity, high-confidence analysis if the given language understanding system generates a single, canonical syntactic parse, finds only one sense of *Jessica* and one sense of *apple* in its lexicon, and can readily disambiguate between multiple senses of *eat* given the fact that only one of them aligns with a human agent and an ingestible theme. At the other end of the complexity and confidence spectrum is the analysis of a long sentence that contains multiple unknown words, results in a fragmented syntactic parse, and yields multiple similarly scoring semantic analyses of unconnected chunks of input.

Systems.   The transition from models to systems moves us yet another step away from the neat and abstract world of theory. Models are dedicated to particular phenomena, while the overall task of agent operation involves integrating the treatment of many phenomena into a single process. Thus, building comprehensive agents requires integrating the computational realizations of the models of individual phenomena. This, in turn, requires managing potential cross-model incompatibilities, a notion we will now unpack.

Any program of research and development must take into account economy of effort. If algorithms and computational systems for certain processing modules of a cognitive agent already exist, then developers should at least consider using them. However, importing code, even well-performing code, comes at a cost. Externally developed components and tools are likely to implement different explicit or implicit models, thus requiring an added integration effort.

Consider an example from the realm of natural language processing. Different language analysis and generation tools available for general use rely on different inventories of parts of speech, syntactic constituents, and semantic dependencies. So, if an off-the-shelf preprocessor or syntactic parser is to be imported into a language understanding system, the form and substance of the primitives in the source and target models must be aligned—which not only requires a significant effort but also might force modifications to the target model.

There is no generalized solution to the problem of cross-model incompatibility since there is no single correct answer for the abstract analysis that underlies cognitive modeling, and people are quite naturally predisposed to hold fast to their individual preferences. So, model alignment is an imperative of developing agent systems that must be proactively managed. However, its cost is significant and involves more than just the initial integration effort since externally developed resources can change their output content and format over time. In fact, the cost of importing processors has strongly influenced our decision to develop a lot of the LEIA models and systems in-house. Still, as discussed in section 2.2, imported, data-driven systems are a natural fit for the flanks of the architecture, Perception Recognition and Action Rendering, and they can also inform certain kinds of reasoning—for example, automatic X-ray analysis can contribute to clinical decision-making. The lesson to be learned is that it is important to assess the costs against the benefits of importing system modules, as well as computational models, into a cognitive agent system.

Another challenge of system building is that all subsystems, be they imported or developed in-house, are error prone. Even the simplest of capabilities, such as part-of-speech tagging, are far from perfect at the current state of the art. This means that downstream components must anticipate the possibility of upstream mistakes and prepare to manage the overall cascading of errors—all of which represents a conceptual distancing from the model that is being implemented.

Because of the abovementioned and other practical considerations, implemented systems are unlikely to precisely mirror the models they implement. This complicates the task of assessing the quality of models. If one were to seek a pure evaluation of a model, the model would have to be tested under the unrealistic precondition that the system implementing the model was provided with perfect upstream results. In that case, any errors would be confidently attributed to the model itself. However, meeting this precondition typically requires human intervention, and introducing such intervention into the process would render the system not truly computational in any interesting or useful sense of the word. The system would amount to a model-system hybrid rather than a computational system. As long as one insists, as we do, that systems be fully automatic, any evaluation will be namely an evaluation of a system, and, in the best case, it will provide useful insights into the quality of the underlying model.

2.4 Microtheories

In LEIA research, we have traditionally called models of the type just described *microtheories*. Microtheories are explanatory, broad-coverage, heuristic-supported treatments of cognitive processes that are implemented in computational systems and enhanced over time. They detail how LEIAs interpret phenomena in language and the world, how they implement their actions, and how their unobservable internal workings are organized. Some noteworthy aspects of microtheories are as follows:

- Microtheories can have a **broad or narrow** purview. An example of a broad microtheory is the one that encapsulates the overall cognitive architecture (section 2.1). An example of a narrow microtheory is the one used to interpret elided verb phrases (section 5.2).

- Broad-purview microtheories **subsume** narrow-purview ones. For example, the agent architecture microtheory subsumes, among others, the microtheories of dialog management, assessing actionability, and explanation. Similarly, the microtheory of natural language processing subsumes the microtheories of language understanding and language generation, each of which, in turn, subsumes microtheories for treating a very large number of linguistic phenomena.

- Microtheories are **not developed in isolation**; they must integrate with others in the cognitive architecture and associated cognitive system.

- Microtheory development involves an **initial top-down descriptive analysis** of the problem space, followed by detailed, algorithm-supported treatments of specific phenomena that are implemented and then continuously enhanced. Top-down analysis helps to avoid the situation, common to demonstration systems, in which solutions that work for a handful of examples must be thrown away each time a new complication is encountered.

- Microtheories have to be **implementable**; they cannot rely on unfulfillable prerequisites or hopes that somebody else will develop methods for meeting the prerequisites. They must, however, allow for the eventuality that, in a given situation, certain kinds of evidence might be incomplete, imperfect, or unavailable.

- Microtheories must specify how to **integrate heuristics** generated by other microtheories or obtained from external knowledge resources. For example, they have to (a) handle conceptual and formalism-based mismatches between imported and native resources, (b) maintain explainability to the degree possible when integrating inherently unexplainable evidence from data-driven systems, and (c) estimate how incorporating unexplainable heuristics affects the agent's overall confidence in the results of its operation.

- Microtheories are developed **over time**, usually starting with the phenomena that are either the most easily treatable or the most urgent for a given application. The solutions leave the door open for future enhancements. Naturally, the problems only get more difficult as one approaches the hardest of cases.
- Agents are configured to **independently assess** whether the current state of a relevant microtheory allows them to act confidently in a given context—for example, whether their coreference procedures can confidently identify the referent for a particular pronoun.
- Based on their confidence estimates, along with features of the situation (e.g., high-stakes or low-stakes) agents are configured to make **actionability judgments**: that is, decisions about whether or not they can proceed to reasoning about action. The methods of making these judgments are recorded in the microtheory of actionability.
- Actionability judgments reflect **self-awareness** and are the key to creating agents that can be useful, trusted collaborators in the near term (they will not just guess in a high-stakes situation), while becoming more sophisticated over time.

When describing LEIAs, the terms *model* and *microtheory* are interchangeable.

2.5 Methodology of Practice: An Accent on System Implementation

Methodology is key to every aspect of content-centric cognitive modeling, and it will be addressed throughout the book. This section focuses on two methodologies aimed at making LEIA development feasible: (1) simpler-first, extensible system development and (2) using graphics and toolsets to optimize developer efforts.

2.5.1 Simpler-First, Extensible System Development
No matter which phenomenon a microtheory addresses, LEIA developers start by asking, "Which cases are *simple*, and which feature values manifest that simplicity?" The simpler cases are implemented right away, with the objective of developing a minimal viable product. The latter then becomes the basis for enhancements over time.

Although simpler-first modeling for LEIAs might sound like the *low-hanging fruit* approach that is common to data-driven AI, these strategies are quite different on two grounds. First, low-hanging fruit systems like those developed for various DARPA-sponsored competitions are typically configured to maximize performance on the task itself and are not extended past the initial limited scope of phenomena and/or coverage.[21] By contrast, simpler-first modeling for LEIAs is part of the development of comprehensive microtheories that must cover all instances of a phenomenon, even if, at a given time, some examples can only be treated in an underspecified manner and/or with low confidence. The second difference between LEIA-oriented simpler-first modeling and data-driven low-hanging-fruit approaches is that LEIAs have metacognition whereas data-driven systems do not. This means that LEIAs are aware of how confident they are in their treatment of each example and why. Confidence estimates enable them to then make decisions about actionability.

Although simpler-first modeling applies to all aspects of agent functioning, its utility for language understanding offers particularly striking examples. Certain types of linguistic metaparameters—such as structural simplicity, parallelism, prefabrication, and ontological typicality—manifest so widely and prominently across the language system that they can serve as a conceptual starting point for simpler-first modeling.[22] So, whether one is building a model of verb-phrase ellipsis resolution, nominal compound interpretation, lexical disambiguation, or new-word learning, one can start by asking: Which kinds of attested occurrences are structurally simple, and which feature values manifest that simplicity? Can syntactic and/or semantic parallelism effects be leveraged in analyzing any of the examples? Can any of the occurrences be treated using prefabricated components, such as lexically or ontologically grounded constructions? Does the analysis of any of the occurrences rely centrally on ontological knowledge—that is, an understanding of how the world typically works? And, finally, do multiple feature values reflecting different metaparameters corroborate the same language-analysis answer?

To concretize the above discussion, consider the following minimal pair of examples, which illustrate the type of verbal ellipsis called gapping.

(2.2) a.   Delilah is studying Spanish and Dana __, French.

b.   ? Delilah is studying Spanish and my car mechanic, who I've been going to for years __, fuel-injection systems.

Gapping is best understood as a construction (a prefabricated unit) that requires the overt elements in each conjunct (the arguments and adjuncts) to be syntactically and semantically parallel.[23] It also requires the sentence to be relatively simple and ontologically typical. The infelicity of (2.2b), indicated by the question mark, results from:

- the lack of simplicity: the second conjunct includes the relative clause *who I've been going to for years*;
- the lack of syntactic parallelism: whereas the second conjunct contains a relative clause, the first does not; and
- the lack of ontological typicality: whereas languages are a typical topic of study, fuel-injection systems are not.

This example illustrates that the metaparameters introduced above are grounded in linguistic reality, which explains why they have proven so useful for modeling quite diverse phenomena—and not only in English but in other languages as well.[24]

As mentioned earlier, in simpler-first modeling, agents must be able to independently determine how different classes of examples are treated by the model and with what confidence. There is no oracle to tell agents that they *can* confidently understand *this* example, while *that* example is too hard. Although it might seem odd to even mention oracles, they have been the cornerstone of many task-specific competitions that have been used to gauge progress in AI, most notably in the field of natural language processing.[25] Task definitions include rule-in and rule-out criteria that are followed during corpus annotation and that constrain what systems are responsible for. Difficult phenomena are manually excluded from purview, which follows the low-hanging fruit principle. As a result, systems that achieve high evaluation scores under such artificial task conditions perform far worse under real-world conditions.[26]

By contrast, LEIAs are geared toward treating all phenomena. For example, their analyses of unexpected (ungrammatical, unknown-word-filled) language inputs might be underspecified and of low confidence, but they may suffice as input to decision-making about how to respond—which might involve asking the speaker for clarification, waiting to see what happens next, attempting to learn unknown expressions on the fly, and so on. This points to the fact that, although simpler-first modeling applies to microtheories individually, agent operation invokes many microtheories simultaneously; and where one microtheory might not suffice—for example, a language input might not be sufficiently understood, another can step in—for example, the agent can ask the speaker for clarification.

Another example of simpler-first modeling comes from the realm of tutoring. In the Maryland Virtual Patient clinician-training application (see section 8.5), a virtual tutor provides guidance related to good clinical practice. For example, if a trainee diagnoses a disease or recommends a test or procedure that is not clinically appropriate, the tutor points out the error. This kind of tutoring support, combined with trial-and-error learning through interacting with virtual patients, results in a useful training environment even though, within the parameters of the project, the tutor was not designed to advise in every way that a human expert could.

Since the evolutionary nature of microtheories is so central to LEIA development, chapter 5 is wholly devoted to illustrating it.

2.5.2 Graphics and Tools

Developing large-scale, extensible knowledge-based systems would be impossible without the support of tools that allow developers to visualize, inspect, and modify knowledge bases, cognitive models, and the results of system processing. Two kinds of graphics and tools are key to this collaboration: diagramming tools for computational cognitive modeling, and the specialized toolkit that supports LEIA development, called DEKADE.

Diagramming tools Computational cognitive modeling is demanding, particularly for comprehensive agent systems whose modules need to interact in complex ways. We have found diagramming using state-of-the-art tools to be an invaluable support for several reasons.[27]

1. Our cognitive models are algorithms not unlike those that have traditionally been represented graphically in the field of computer science.

2. Modern diagramming environments are engaging and produce pleasing results, which eases the cognitive load when one is trying to solve difficult modeling problems.

3. Cognitive system development requires close collaboration between knowledge engineers and system engineers. It is not sufficient for knowledge engineers to pass off algorithms to system engineers in a unidirectional pipeline because knowledge engineers understand how the system will evolve over time, which can have important implications

for software design. Diagrams help to establish the common ground between knowledge engineers and system engineers, defined as the level of an algorithm that they both need to fully understand and sign off on. Below the common ground, system engineers can implement as they see fit. The common ground is actually deeper than one might expect: to develop cognitive systems that will stand the test of time, knowledge engineers and system engineers both need to stretch beyond their traditional roles and comfort zones.

4. It is important to be able to explain cognitive models to the outside world, particularly to domain experts who will want to ensure their veracity. Diagrams help to establish a content-verification level of algorithms for this purpose.

The DEKADE environment   DEKADE is the graphical Development, Evaluation, Knowledge Acquisition, and Demonstration Environment that supports LEIA research and development. It enables:

- viewing and acquiring the lexicon, ontology, and episodic memory;
- running the natural language understanding engine in test-debug mode; this involves inputting texts, having them semantically analyzed, viewing their scored TMR candidates, viewing traces of processing, and amending the knowledge and processors as needed;
- running the natural language generation engine in test-debug mode; this involves selecting or generating Generation Meaning Representations (GMRs) as input, running the generator, viewing scored sentence outputs, viewing traces of processing, and amending the knowledge and processors as needed; and
- viewing traces of the agent's reasoning during applications using dynamically populated under-the-hood panels.

Screenshots and additional details about DEKADE functionalities are available in the online appendix.

For purposes of this discussion, the important point is that we are in a wholly different technological world from the early days of knowledge-based AI over a half century ago, when everything about computation was slow, clunky, and limited. Today, powerful, convenient tools can be built fast, enabling the development of systems that rely on large, high-quality, human-curated knowledge bases.

2.6   Recap of Content-Centric Cognitive Modeling

Content-centric cognitive modeling offers a path from theories to application systems and pursues near-, mid-, and long-term objectives. Among the near-term objectives is to configure useful systems across domains while maximally reusing knowledge bases and agent functionalities. Among the mid-term objectives is to reach the critical mass of high-quality ontological and lexical knowledge to support the agent's independent lifelong learning. Among the long-term objectives is to operationalize high-quality lifelong learning within ever more sophisticated LEIA applications. In this paradigm, knowledge that supports a system—be it a demonstration system or a deployed one—is not thrown away; it is added to the cumulative knowledge in the underlying content-centric cognitive model and can be used in later systems.

2.7   Comparisons with Other Approaches

Contributions to the field of artificial intelligence can be described according to many parameters, such as method of computation, method- or task-driven application selection, target application, timeframe for the development effort, balance of research and development, presence or lack of human-inspired modeling, explainability, extensibility, and so on. LEIAs differ from data-driven AI systems with respect to most of these parameters, and they differ from other cognitive systems with respect to key ones. The following subsections point out some salient contrasts.

2.7.1   Thumbnail Juxtaposition with Data-Driven AI

The majority of modern AI systems use data-driven methods to configure standalone applications that can be deployed in the near term. The field is largely driven by commercial interests, which prioritize configuring immediately useful technologies. However, scientific aims are also pursued, primarily in the academy and through governmental organizations. As new methods of computation gain the field's interest—most recently, deep learning—practitioners seek application areas that can be served by them.

This book is not the place for a survey of AI. First, the book's contribution lies elsewhere: presenting an alternative (non-mainstream) path toward configuring systems that display aspects of human intelligence. Second, it is impossible to adequately survey a field as broad as AI in a short space. And finally, surveys of subfields and individual research thrusts are readily available.[28] Rather than anything survey-like, we present a thumbnail sketch that orients around four points of contrast between data-driven AI and LEIAs. The contrasts involve meaning, reasoning, applications, and explainability.

Meaning in data-driven AI   Data-driven AI operates over uninterpreted words; it does not pursue the computation of meaning as understood by linguists and philosophers. When the terms *meaning* and *semantics* are used to describe data-driven systems, they are redefined according to what the technologies can actually supply.

Although data-driven systems do not compute full semantic and pragmatic analyses, significant effort has gone into computing semantic and pragmatic features that can improve system performance, such as semantic roles, coreference relations, temporal relations, and discourse relations.[29] In addition, the statistical approach called *distributional semantics* operationalizes the intuitions that "a word is characterized by the company it keeps" (Firth, 1957) and "words that occur in similar contexts tend to have similar meanings" (Turney & Pantel, 2010).[30] Distributional models are good at computing similarities between words. For example, they can establish that *cat* and *dog* are more similar to each other than either of these is to *airplane*, since *cat* and *dog* frequently co-occur with many of the same words: *fur, run, owner, play*. Moreover, statistical techniques like Pointwise Mutual Information can be used to detect that some words are more indicative of a word's meaning than others. For example, whereas *fur* is characteristic of dogs, very frequent words like *the* or *has*, which often appear in texts with the word *dog*, are not especially characteristic of dogs. Although distributional semantics has proven useful for applications like document retrieval, it is not a comprehensive approach to computing meaning since it only considers the co-occurrence of words. Among the things it does not consider are:

- the ordering of the words, which can have profound semantic implications: *X attacked Y* versus *Y attacked X*;
- their compositionality, which is the extent to which the meaning of a group of words can be predicted by the meanings of each of the component words; for example, in most contexts, *The old man kicked the bucket* does not mean that he struck a cylindrical open container with his foot;[31] and

- any of the hidden sources of meaning in language, such as ellipsis and implicature.

Reasoning in data-driven AI   Since data-driven AI systems do not compute meaning, they cannot reason in humanlike ways, even though some create the illusion of reasoning—as when playing chess or doing machine translation. But what defines reasoning for machines? Some developers believe that the definition lies in benchmarks.[32] That is, if one establishes a criterion for reasoning, and if a system fulfills that criterion, then it has reasoned.

At the time of writing, the assessment of reasoning is being rigorously discussed with respect to large language models (LLMs), which can generate text and respond to certain kinds of queries in humanlike ways.[33] Responses to this new technology by professionals and lay people abound.[34] With respect to LEIAs, two points about LLMs are important:

1. **LEIA development is not superseded or invalidated by LLM technologies.** Mahowald et al. (2023) explain LLM achievements and deficiencies in terms of the juxtaposition between (a) formal linguistic competence, which is the ability to generate text that sounds like the given language, and (b) functional competence, which is the ability to understand and use language in the world. They say, "LLMs show impressive (although imperfect) performance on tasks requiring formal linguistic competence, but fail on many tests requiring functional competence" (p. 1). Bubeck et al. (2023, pp. 93–94) list deficiencies of the LLM called GPT4: the model does not know when it is guessing; it makes up facts; it uses a very limited context; it cannot learn or update to changes in the environment; it cannot be personalized to users; it appears to suffer from cognitive biases in reasoning; and its post hoc explanations can range from useful to wrong or unrelated. So, while LLMs are undoubtedly an exciting breakthrough that promises significant technological advances in a variety of practical applications, they do not offer theoretical advances in understanding human cognition or in creating computational models of reasoning based on understanding. Instead, they generate plausible texts by computing the most probable continuation of a text on the basis of an enormous store of text examples.

2. **LLMs can serve LEIAs in similar ways as other data-driven tools.** LLM technology can and should be used to support computational cognitive

modeling. The formal, surface-level linguistic competence of LLMs opens up possibilities for multi-engine approaches to certain tasks undertaken by LEIAs. So far, we have experimented with their use in selecting among candidate sentences in language generation (section 4.3) and in preparing learning materials for LEIAs (section 7.1.3). It is very early days in the availability of LLMs, so we have only just begun to flesh out their potential utility for LEIA operation.

As Yogi Berra opined, it is difficult to make predictions, especially about the future. At present, deep reasoned assessments of the impact of deep neural network–based LLM technology on science are clearly premature. We intend to return to this topic in our future work.

Let us return to the matter of comparing reasoning by LEIAs with reasoning by data-driven systems such as LLMs. Data-driven systems do not reason in humanlike ways, so when they are asked to explain, they have to concoct something post hoc, which might be useful, wrong, or irrelevant.

Application areas Most data-driven AI applications address a single capability and can be referred to as silo systems. Readers are well acquainted with the successes of silo AI, which offers systems for web search, route guidance, object recognition, playing chess, making product recommendations, machine translation, and so on.

But while silo AI has spectacularly fulfilled some desiderata, others remain outstanding. For example, silo systems tend to suffer from rigidity. If an AI system is trained to play a game on a grid with a particular layout, if any aspect of that layout is changed, then the system needs to be retrained from scratch—one cannot just tell the system, "Play just as before but use the additional rows as well." Another concern involves ceilings of results. Leaderboards in system competitions regularly show the winner's score in the 70th or 80th percentile, at which point developers abandon the task because the preferred methods offer no inroads for improvement. Yet another issue that has not made it onto the agenda of mainstream AI is addressing the challenges of complex application areas, such as clinical medicine viewed holistically. Finally, the lack of explanatory power of data-driven AI has been recognized as a serious limitation to its adoption by end users in critical domains, which is the topic to which we now turn.

Explanation in data-driven AI Explanation is a problem for data-driven AI systems because their methods of computation are opaque even to their developers.[35] For this reason, they are regularly described as *black boxes*. Their lack of explainability is problematic because, in a recent survey, explainability ranked third highest among the ten major risk factors potentially preventing the broad acceptance and deployment of AI, following cybersecurity and regulatory compliance (McKinsey Analytics 2021). Why does explainability matter? Because in domains like health care, law, defense, and finance, outcomes carry consequences, and neither the decision-makers responsible for those consequences nor the individuals affected by them are prepared to blindly trust a machine.

Curiously, despite the absence of explanatory power, the production and approval of AI applications has been steadily accelerating. As of June 2021, the FDA cleared 343 AI-ML-based medical devices, with over 80 percent of clearances occurring after 2018 (Matzkin, 2021). The supply of new AI systems has continued unabated even though their adoption has been less than enthusiastic. Fully 70 percent of these devices offer radiological diagnostics and typically claim to exceed the precision and efficiency of humans. But, according to Gary Marcus (2022), as of March 2022, "not a single radiologist has been replaced." So, regulators keep approving systems whose operation cannot be explained, and developers keep hoping that their systems, though unexplainable, will be adopted. Why?

One reason is that, historically, many technologies have been adopted before their principles of operation were fully understood. For example, the steam engine was invented well before the laws of thermodynamics were formulated, and Röntgen named his discovery "X-rays," where the "X" signified "unknown." In the absence of causal explanations, trust in novel technologies may be earned on the basis of correlational and statistical evidence. In other words, randomized control trials can stand in lieu of explanation as sufficient grounds for adopting a technology—as was the case with the FDA approvals of the latest batch of AI-grounded medical devices.

Another reason why unexplainable technologies can be broadly adopted is that users are willing to accept inconsistent quality of system output. Google Translate is a good example. It has become a staple of everyday life despite showing big differences in error rates for different language pairs, topics, text genres, and stylistic registers.[36] If a reliable, high-quality translation is required, results from Google Translate must be postedited by a human expert.

While most people do not know or care how their microwave ovens, GPS systems, or online translators work (cf. Gray, 2019), they *do* want to know why a particular medical treatment, investment strategy, or military command decision is being endorsed. They also want to know whether the recommendation comes from a live expert, an AI system, or a collaboration between the two. Current AI systems cannot provide such explanations.

Issues related to the adoption and explainability of AI are of wide interest to developers, consumers, regulatory agencies, funding agencies, foundations, venture capitalists, government and corporate decision-makers, educators, and media influencers. While opinions and decisions in this realm are of vital importance to the future of the AI enterprise, here we concentrate on the supply side of the issue: the ways in which the developers of data-driven AI systems are addressing explainability.

The need for explanation is fully recognized by developers of data-driven AI systems. The past decade saw a wave of research projects devoted to explainability, spearheaded by DARPA's Explainable AI (XAI) program.[37] This research program concentrated on providing explanations for AI applications that are based in deep learning. Deep learning has been a popular method of late because it dispenses with the use of externally defined features. On the positive side, this avoids the expensive human data preparation required by supervised machine learning. (Even though, according to *The Economist*, as of January 2020, data preparation still claimed over three quarters of the time allocated to machine learning projects.[38]) On the negative side, the uninterpretability of features makes deep learning–based systems even less explainable than their predecessors.

The way the XAI community has attempted to square this circle is by redefining explanation. XAI research does not seek to explain how systems arrived at their output. Instead, it has concentrated on "*post hoc* algorithmically generated rationales of black-box predictions, which are not necessarily the actual reasons behind those predictions or related causally to them . . . [and which] are unlikely to contribute to our understanding of [a system's] inner workings" (Babic et al., 2021). The XAI literature has "largely forgotten" about the goal of explaining what is going on inside the black boxes of machine learning–based AI, even though that is the key to trust and adoption by end users (*The Economist*).[39] In short, XAI-style explanations may benefit the research community, but they do not serve system users.

The need to tailor explanations to different categories of users and purposes has led to the emergence of the human-centered explainable AI (HCXAI) research community.[40] This community is developing a variety of classification schemes addressing issues like:

- the explanation needs of various users;
- the purposes of explanations; these are called "XAI goals" in Barredo Arrieta et al. (2020), and they include trustworthiness, causality, transferability, informativeness, confidence, fairness, accessibility, interactivity, and privacy awareness;
- the inventory and needs of applications requiring explainability, such as decision support and model fairness evaluation;
- evaluation methods for HCXAI systems; and
- the typology of explanation-seeking questions expected from different stakeholders.

HCXAI work is in the tradition of human factors and human-computer interaction (HCI) research, so a typical contribution is a description of a user-centered method for designing question-driven explanation systems.[41] The HCXAI community is also making an effort to involve experts from different fields, including HCI, psychology, machine learning, and the social sciences (Ehsan et al., 2022).

So far, the HCXAI community has mostly proposed adaptations to extant approaches to explanation (Liao & Varshney, 2022), reflecting a commitment to data-driven AI that includes:

- reliance on the familiar data-driven solutions, despite their unintepretability;
- a focus on task competitions with standard numerical evaluation criteria and leaderboards; and
- the acceptance of partial success (such as the best system correctly answering 81.1 percent of the questions) without the need to address how results can—or, more often, cannot—be improved if the selected methods were to continue being developed.

HCXAI explanations can, for example, rely on a predefined list of eventualities or situations in the application domain, each of which is linked to an explanatory text. Then, when the system classifies a state of affairs as one of such eventualities or situations, it can display the corresponding canned text in lieu of an explanation. We hypothesize that explanations of this genre will be deemed acceptable only for applications where explanations are sought merely as a matter of curiosity, as in entertainment applications, not as a matter of importance.

The next step toward increasing the societal impact of AI is to develop systems that function as full-fledged members of human-AI teams. Such systems will, for example, provide suggestions, explanations, and reminders to human experts, thus lowering their cognitive load and helping them to avoid errors and omissions. They will take advantage of high-quality, silo-style AI while necessarily supplementing it with many additional capabilities. Such systems have been referred to as *orthotic, human-in-the-loop AI*, or simply *human-AI systems*.[42] LEIAs are one example of such systems, but they do not stand alone. Developers of cognitive systems and their underlying cognitive architectures have been working on this problem for decades.

2.7.2 Typical Choices in Cognitive Systems Research

Cognitive systems developers build application systems that are grounded in a cognitive architecture.[43] Cognitive functionalities are typically embedded in simulated or robotic systems that are limited to a narrow domain and set of functionalities. The limitations of size and scope are justified by the considerable complexity of not only implementing the individual capabilities (vision processing, language understanding, decision-making, and so on) but also of integrating them into an overarching system. As Laird (2012) says, "There are tradeoffs between the amount of knowledge that can be stored about a situation, the accuracy with which it can be stored, and the efficiency and accuracy with which it can be retrieved in the future" (p. 32).

Although cognitive systems developers are working toward human-level AI, it remains a long-term goal. So, development efforts must be both amply supported by research and commensurate with the available workforce and time constraints. This has resulted in decisions to distribute labor in various ways. Some developers work on a single module of a comprehensive AI system, often assuming the availability of unattainable prerequisites and envisioning future system integration of an unspecified nature. Other developers work in a single small domain, addressing multiple modules and their interactions but not the various preconditions for transcending their domain of choice. Still others work on a single application, focusing on its utility without concern for its contribution, if any, to developing comprehensive AI systems. There are both scientific and practical justifications for organizing research and development in these ways. However, if we are to make progress in the long game of human-level AI, these simplifying methodologies cannot stand forever. The reason why is that the solutions to simplified problems are not additive. Imagining that they *are* would be like saying that skyscraper technology equals tent technology plus log cabin technology plus two-story suburban home technology.

Our insistence that large, deep, heterogenous, agent-appropriate knowledge must be a core consideration from the outset of any agent-development effort does not mean that we are unaware of, or unsympathetic to, the circumstances that lead to developing narrow-coverage cognitive systems. Developers of such systems have important engineering concerns, such as achieving near-real-time performance, dealing with format conversions between the inputs and outputs of system modules, and achieving satisfactory performance of the individual modules themselves. Indeed, the complexity of integrating the diverse processing modules into a comprehensive cognitive system application is formidable even when knowledge coverage is limited.[44] Moreover, we are all subject to extra-scientific concerns, such as the funding climate, which does not reward work on building high-quality knowledge bases. Still, it is methodologically unsound to invoke, *a priori*, questionable simplifications without seriously assessing how they impact moving from demo systems to deployed ones.

To summarize, cognitive systems typically use knowledge bases that are small and shallow, providing just enough to cover the needs of a particular application.[45] The goal is to reduce the complexity of system engineering and improve system performance. However, while this approach fosters the development of certain individual AI components, it hobbles the necessary work on overarching, integrative, long-term challenges.

2.7.3 Cognitive Architecture Research

Cognitive architecture researchers propose a typology of, and requirements for, the knowledge components of agent systems, and they recommend formalisms for, and approaches to, processing by such agents. These recommendations are then implemented by cognitive systems developers. Naturally, some researchers wear both hats, inventing their own cognitive architectures and building systems atop them. LEIA research falls into this category. No matter the source of the architecture, be it in-house or imported, the more theoretically motivated application systems attempt to carry out as little "tailoring" (to use Forbus's [2019, p. 21] term) as possible—that is, they try to maintain the theoretical principles of the cognitive architecture while still meeting application requirements.

Cognitive architectures reflect theories about how the human mind works and hypotheses about how they can be operationalized in computer systems. The survey of cognitive architectures in Langley, Laird, and Rogers (2009) describes nine capabilities that any good cognitive architecture must have (1) recognition and categorization, (2) decision-making and choice, (3) perception and situation assessment, (4) prediction and monitoring, (5) problem solving and planning, (6) reasoning and belief maintenance, (7) execution and action, (8) interaction and communication, and (9) remembering, reflection, and learning.

Many cognitive architectures have been proposed and are under development.[46] Whereas in some cases, their differences reflect different theories about cognition, in other cases, they reflect differences in the priorities of their developers and/or differences of opinion about how AI will advance over time. One example involves language processing. Many cognitive architectures treat language understanding as if it were an isolated module. This dramatically simplifies reality. To interpret what somebody means, one might need to see and hear what's going on in the context, reason about a joint plan, and/or understand the speaker's goal. Designers of cognitive architectures know this, but, if the focus of their research lies elsewhere, then they can justify language-oriented simplifications as a steppingstone.[47]

However, in our opinion, theories of cognitive functionalities must, from the outset, be informed by a realistic reckoning of the actual scope of eventualities that must be covered. Awareness of these eventualities is a core aspect of the knowledge that must be recorded in any cognitive system that implements a given cognitive architecture. Adopting too many simplifying assumptions too early runs the risk of needing to summarily discard and redesign components, or even the overall architecture, if the systems using it are ever to advance beyond small domains.

2.7.4   The Main Takeaway from These Comparisons

While all of the work described in this section makes important contributions to AI, none of it addresses what we consider a core and pressing issue of explainable intelligent behavior: compiling broad-scale, sophisticated *knowledge content* that takes a long view—that is, that foresees a wide range of interconnected agent capabilities across domains and applications.[48] Cognitive architectures research might be absolved of this since, as a theoretical enterprise, it is responsible only for accommodating different types of knowledge, not for acquiring it. However, we believe that the cognitive systems community should be more concerned than it currently is about non-toy knowledge because acquiring and manipulating it is a prerequisite to scalability—that is, to advancing from prototype systems in a single domain to full-scale, deployable systems across domains.[49] This is the niche that LEIAs seek to occupy, in ways described in the remainder of this book.

3 Knowledge Bases

A distinguishing feature of LEIAs is their reliance on a large amount of stored knowledge. Their core knowledge bases are the ontology; the lexicon; lexicon analogues for nonlinguistic channels of perception, such as an opticon for the interpretation of visual stimuli; and episodic memory. The theoretical and methodological principles underlying the content and organization of these knowledge bases are detailed in Nirenburg and Raskin (2004) and McShane and Nirenburg (2021). This chapter recaps the basics and describes how two microtheories—involving complex properties and scripts—have recently evolved to account for the demands of learning and explaining by LEIAs. But first it is important to show why well-known, large "knowledge" resources are not sufficient.

3.1 Why Preexisting Resources Don't Fill the Bill

There are many human-curated repositories of information about language and the world—lexicons, thesauri, wordnets, grammars, ontologies—but they do not contain knowledge in a LEIA's sense of the word and, therefore, are not directly useful to them. This is unfortunate because they reflect a lot of human analysis, and one cannot help but feel that they should be more useful for developing intelligent systems.[1]

Human-oriented lexicons When human-oriented lexicons were digitized in the 1980s, there was a surge of interest in automatically converting them into machine-oriented knowledge bases.[2] Developers expected that systems could learn an ontological subsumption hierarchy from the hypernyms that introduce most dictionary definitions (*a dog is a domesticated carnivorous mammal*) and extract other salient properties as well (*. . . that typically has a long snout*). But there were snags:

1. Senses are often split too finely for even a person to understand why.
2. Definitions regularly contain ambiguous words or idiomatic expressions.
3. Sense discrimination is often left to examples, meaning that the user must infer the generalization illustrated by the example.
4. The hypernym that typically begins a definition can be of any level of specificity—*a dog is a(n) animal / carnivore / domesticated carnivore / mammal*—which confounds the automatic learning of a semantic hierarchy.
5. The choice of what counts as a salient descriptor is variable across entries: *a dog is a domesticated carnivorous mammal; a turtle is a slow-moving reptile.*
6. Definitions can be circular: *a tool is an implement; an implement is a tool.*

After more than a decade's work attempting to automatically adapt machine-readable dictionaries for use in natural language processing, by the early 1990s, that community concluded that this line of research had little direct utility: machine-readable dictionaries simply required too much human-level interpretation to be useful to machines in the ways originally hoped (Ide & Véronis, 1993). Unfortunately, these problems do not go away even if we ask LEIAs to semantically analyze the definitions. It's a chicken-and-egg problem: LEIAs need to learn the very kind of knowledge that is needed to disambiguate the dictionary's descriptions.

Thesauri   Most thesauri list clusters of words without explaining what distinguishes them, their main use being to remind native speakers of a word they could not recall. If you have ever tried to use a thesaurus for a language you don't know very well, you realize the problem facing agent systems. Moreover, thesauri can group semantically diverse meanings together—not only synonyms and near-synonyms but also rather distant hyponyms and troponyms. There do exist explanatory thesauri, such as Hayakawa (1994), which provide explanations of the distinctions between word usages; however, these descriptions present the same ambiguity challenges as the definitions in standard dictionaries.[3]

WordNets   The original WordNet (Miller, 1995), from which wordnets in other languages followed suit, is a lexical database of English whose initial goal was to model human lexical knowledge. It is organized as a semantic network of four directed acyclic graphs, one for each of the major parts of speech: noun, verb, adjective, and adverb. Words are grouped into sets of synonyms called synsets. Synsets within a part-of-speech network are connected by a small number of relations. For nouns, the main ones are subsumption ("is a") and meronymy ("has as part"); for adjectives, antonymy; and for verbs, troponymy, which is subsumption involving the manner of the action.

Although WordNet was not originally developed for computational aims, it has been widely used by the natural language processing community for a similar reason as machine-readable dictionaries: It is large, containing 155,287 unique strings and 206,941 word-sense pairs, and it is freely available.[4] However, as a potential resource for learning by LEIAs, WordNet is as unwieldy as regular large dictionaries:

- Its definitions are in plain English, so they can be ambiguous or idiomatic.
- It exhaustively lists attested word usages, no matter how rare or narrowly applicable. For example, there are ten senses of *heart*, ten of *cat* (eight nominal and two verbal), and eight of *beaver* (seven nominal and one verbal).[5] Although LEIAs are designed to treat lexical ambiguity, these large inventories, which include rare senses, present a serious obstacle for making practical progress on computational semantics.
- The relative frequency of senses is not indicated. For example, two of WordNet's nominal senses of *dog* are "a hinged catch that fits into a notch of a ratchet" and "metal supports of logs in a fireplace." Similarly, there are two verbal senses of *cat* meaning to beat with a cat-o'-nine-tails and to vomit. Even an impressionistic indication of rare or specialist-domain senses would have increased the potential utility of this resource for LEIA learning.
- Multiword expressions are listed as if they were regular senses of one of their constituent words, with no indication that the entire collocation is needed to convey the meaning. Examples are *to play house* (listed as a sense of *house*), *to have a change of heart* (listed as a sense of *heart*), and *to do something by the book* (listed as a sense of *book*).
- Syntactic and semantic information about the arguments of verbs is not provided, which means that, even in the best case, a LEIA could only learn which verbs might be synonyms or troponyms of other verbs, not all of the dependency-based information needed to process them.
- The results of productive linguistic processes are recorded as word senses, which runs counter to a LEIA's model of human language processing,

which distinguishes stored knowledge from productive processes. For example:

– Metonymies can be listed as regular word senses. For example, one sense of *house* refers to the family living there, in contexts like "I waited until the whole house was asleep." But any object can be used metonymically to refer to someone associated with it: *The backpack <unicycle, blue shirt> just waved at us.*

– Regular personifications can be listed as senses. For example, there is a separate sense of *teacher* whose examples include "books were his teachers" and "experience is a demanding teacher."

The problem with listing the results of productive linguistic processes as word senses in a LEIA's lexicon is that every time the agent encountered the given word, it would have to consider all of the listed senses. For example, if the lexicon contained a sense of *house* meaning "the people in a house," then every input with *house* would result in a candidate analysis using this interpretation. Not only is this hardly likely to model human lexical storage but it also unnecessarily complicates language analysis. A much better solution to treating productive linguistic processes is the one that LEIAs actually use: they consider nonliteral meanings of words only if the senses recorded in the lexicon do not semantically fit the context. For example, since a house cannot be asleep, when a LEIA analyzes *I waited until the whole house was asleep*, it will engage in recovery procedures that consider, among other things, the possibility that *house* is being used metonymically. This is a normal part of its language understanding process (cf. section 4.2.1).

• The classification of verbs is often imprecise. For example, all of the following verbs are considered to be troponyms of *kill* but they do not actually mean *to kill*—they indicate methods of injuring someone that may or may not result in death: *poison, stone, brain, impale, shed blood, electrocute, flight (to shoot a bird in flight), pick off, shoot, saber, tomahawk, strangle.*

• Definitions are often idiomatic or include vocabulary that is more complicated than the word being described. In the following examples, the tricky parts are italicized:[6]

– lynch: kill *without legal sanction*

– murder, slay, hit, dispatch, bump off, off, polish off, remove: kill intentionally and *with premeditation*

- burke: murder *without leaving a trace on* the body
- execute, put to death: kill *as a means of socially sanctioned punishment*
- neutralize, neutralise, liquidate, waste, knock off, do in: *get rid of* (someone who may be a threat) by killing

- Notes about semantic constraints do not specify which participant is being referred to. For example, from the definition of *assassinate*—"murder; especially of socially prominent persons"—one must figure out that the socially prominent people are the theme, not the agents, of the action.

Like any resource, WordNet reflects a large number of choices by its developers, who had particular goals and priorities in mind—none of which was to support learning by intelligent agents like LEIAs. So, our assessment is not of WordNet in the abstract. Instead, the question for us is whether this resource, which reflects a significant societal investment, can contribute to LEIA development. The answer is "yes," but not as a source of automatic learning by LEIAs. Instead, it can be used to jog the memories of knowledge engineers, who can skim it for useful content.

FrameNet   FrameNet is a lexical knowledge base inspired by the theory of frame semantics (Fillmore & Baker, 2009), which is a precursor of construction grammar (Hoffmann & Trousdale, 2013). Frame semantics suggests that the meaning of many words and multiword expressions is best described using semantic frames that indicate a type of event and the types of entities that participate in it. For example, an Apply_heat event involves a Cook, Food, and a Heating_instrument. A frame thus described can be evoked by particular *lexical units* (i.e., words and phrases), such as *fry* and *bake*. FrameNet includes frame descriptions, associated lexical units, and annotated sentences featuring those lexical units. Frame semantics captures observations about language and meaning similar to those made by the theory of Ontological Semantics that underpins LEIAs' language processing. However, frame semantics mixes lexical and ontological knowledge in a way that diverges from our approach.

Because of the high lexicographic quality of FrameNet, we tried to make use of it—specifically, to give LEIAs practice in (a) learning new words and multiword expressions and (b) analyzing sentences from the open domain. Although the experiment that explored this potential was not as fruitful as we had hoped, it was instructive, helping us to clarify what constitutes the kind of learning material that will help LEIAs to walk before they can run.

In order for LEIAs to use FrameNet as a resource, the first requirement is to automatically align FrameNet frames with concepts in the LEIA's ontology. An alignment is hypothesized if two or more FrameNet lexical units are attested in the LEIA's lexicon and mapped to the same or proximate concepts in the same line of inheritance in the LEIA's ontology. According to this heuristic, FrameNet's Ingestion frame aligns with the LEIA's concept INGEST based on the evidence summarized in table 3.1.

For each concept-level alignment, two opportunities open up: the LEIA can learn new lexemes and analyze annotated sentences.

Learning new lexemes   FrameNet frames contain words and multiword expressions that the LEIA does not yet know and can learn—such as the verbs *down, feed, gobble* and *guzzle* and the noun *gulp* in table 3.1. Besides the concept mapping, LEIAs need to learn the syntactic dependency structures in which each word can participate, which is illustrated by the frame's examples. For instance, the verb *gobble* can take a direct object ("Don't gobble yer food so fast") or it can be used with various particles in phrasal verb constructions ("I *gobbled* them *down*," "This year . . . four tons of fresh strawberries will be *gobbled up*"). Similarly, the noun *gulp*

Table 3.1
A sampling of alignments between FrameNet and a LEIA's ontology and lexicon.

FrameNet concept **Ingestion**	LEIA concept INGEST
breakfast.v	breakfast-v1
consume.v	consume-v1
devour.v	devour-v1
dine.v	dine-v1
down.v	–
drink.v	drink-v1
eat.v	eat-v1
feast.v	feast-v1
feed.v	–
gobble.v	–
gulp.n	–
gulp.v	gulp-v1
guzzle.v	–
have.v	have-v5

can take a prepositional phrase with *of* that indicates what is drunk ("She drank a good *gulp* *of whiskey*").

Analyzing sentences that use the word in the known meaning   One of the main offerings of FrameNet is its large repository of annotated sentences, which can be used to test the quality of language understanding by LEIAs. These annotations provide (a) a key lexical disambiguation decision—that is, the meaning of the word whose use is illustrated by the example, and (b) the fillers of that word's case roles. The FrameNet examples can, in principle, also support lexicon learning through bootstrapping. For example, one of FrameNet's examples of *drink* is "They drank hot sake from tiny porcelain cups, . . . ," from which the LEIA can learn that *sake* is a type of BEVERAGE (to understand how it does this, see chapter 7).

Naturally, we manually analyzed FrameNet to some degree before designing our learning experiment, and it was our analysis of frames like the following that gave us reason to believe that the approach might work: Absorb_heat, Arrest, Bragging, Cogitation, Communication_manner, Ingestion, Self_motion. However, when we launched the program on FrameNet overall, the results were much less clean than we had expected. Reasons include the following:

- FrameNet uses both generic and frame-specific case role labels, whereas a LEIA's knowledge bases use only generic ones. For example, FrameNet's Ingestion frame includes an Ingestor and Ingestibles as case roles, whereas a LEIA's concept INGEST uses a generic AGENT and THEME. So, LEIAs have to hypothesize the FrameNet-to-LEIA alignment in each case.

- Some FrameNet frames group entities in ways that our semantic theory does not permit, as by bunching words and their antonyms. For example, the Accompaniment frame includes the words *alone* and *together*. LEIAs cannot automatically distinguish synonyms from antonyms, at least not without consulting additional resources.

- FrameNet frames can cover broader semantic territory than the LEIA concepts to which associated words would link. For example, the frame Cause_to_experience covers *amuse* and *entertain* as well as *terrorize* and *torment*.

- FrameNet bunches literal and metaphorical uses, which is something we strictly avoid. For example, the FrameNet lexical unit devour.v belongs to the frame Ingestion but includes the examples, "On rainy days he devoured books, . . ." and "The houses looked like shambling tents of black straw, their terraces devoured by the glutton of rot . . ."

- Generalizing from the last point, since FrameNet developers were not orienting around the needs of agent systems, some of the examples they chose to annotate are worst cases for automatic analysis. Given the example "We're not talking about children eating deadly nightshade," a LEIA has no way of knowing that *deadly nightshade* should not be learned as an INGESTIBLE (the LEIA's ontology asserts that THEMES of INGEST must be INGESTIBLES). Of course, the agent could be configured to carry out additional work to vet every candidate learnable; however, that kind of work is exactly what we were trying to avoid by using a curated resource like FrameNet as a source for learning.

- All of FrameNet's definitions are in plain English, thus presenting all of the ambiguity challenges of the definitions in human-oriented lexicons and WordNet. For example, *amble* (part of the Self_motion frame) is described as "walk or move at a leisurely pace"; and *bop* is described as "to go quickly or unceremoniously; shuffle along as if to bop music." A LEIA could not automatically learn how these words differ from the core meaning of the frame without knowing the meanings of *leisurely pace*, *unceremoniously*, and *bop music*.

- FrameNet's example annotations provide less added value to LEIAs than they might to other computer systems because our language understanding system identifies case roles as part of its normal operation.

In the spirit of not giving up on using existing resources, we could manually prune FrameNet in order to make it more useful to LEIAs. This would involve removing all metaphorical and unreasonably difficult examples, splitting or removing frames that bunch antonyms, reformulating definitions to be more useful, and the like. We could also try to automate the pruning, at least partially. However, this would make sense only if the methods developed had broader applicability—that is, if they were able to detect learning-suitable material in any text repository. FrameNet is too small to make narrowly focused manual work cost-effective.

To reiterate, the reason we spent effort on exploring the potential utility of FrameNet is because we thought it might offer LEIAs useful practice in learning new words and analyzing open-domain sentences in a setup that included hints based on manual curation. However, since no small amount of manual work is needed to prepare for such a process, further exploration of this resource goes on the list of potential but not imminent knowledge-acquisition methodologies. Our experimentation with FrameNet underscores our main claim about knowledge acquisition for LEIA-like intelligent systems: it is best done in conjunction with system-building. Even with optimal bootstrapping knowledge, automatic semantic analysis and learning are challenging. With resources that are not entirely suited to the task, *challenging* quickly morphs into impossible.

Cyc Cyc is one of the oldest ontology-building efforts to date, described by its developers as a "high-risk high-labor long-term project" (Lenat et al., 1990). Its project leader, Doug Lenat, started the project with the goal of recording a sufficient amount of commonsense knowledge to support any task requiring AI. He writes: "[F]or the last 35 years that Manhattan-Project-like effort has occupied a team of over a hundred knowledge engineers (whom I dubbed 'ontologists' back then)—that's millions of person-hours of writing and testing and debugging IF/THEN rules" (Lenat, 2019). Although initially configured using the frame-like architecture typical of most ontologies, the knowledge representation strategy shifted to a "sea of logical assertions," such that each assertion is equally about each of the terms used in it (Mahesh et al., 1996, p. 21).

In a published debate with Lenat (Lenat et al., 1995), George Miller articulates some of the controversial assumptions of the Cyc approach: that commonsense knowledge is propositional; that a large but finite number of factual assertions, supplemented by machine learning of an as-yet undetermined type, can cover all necessary commonsense knowledge; that generative devices are unnecessary; and that a single inventory of commonsense knowledge can be compiled to suit any and all AI applications. Additional points of concern include how people can be expected to manipulate—find, keep track of, detect lacunae in—a knowledge base containing millions of assertions, and the ever-present problem of lexical ambiguity. Yuret (1996) offers a fair-minded explanatory review of Cyc in the context of AI.

Although we give Lenat and the ontologists at Cyc a lot of credit for taking on the challenge of building such a large knowledge base, we do not use it. As previously mentioned, our experience shows that knowledge bases and processors need to be developed together. We do not exclude that Cyc might be useful in some way as LEIAs evolve over time, but assessing how it might be employed would be a large program of work in itself.

It is plain common sense that reusing existing resources has the potential to save effort. It was, therefore, important to spend time here describing our lessons learned from attempting to use available knowledge resources and explaining why the current offerings do not directly fulfill the needs of LEIAs. Now we will turn to the resources that LEIAs actually need in order to implement the computational cognitive models underlying them.

3.2 Ontology

A computational ontology is a model of the world designed to foster reasoning by agent systems. The ontology used by LEIAs is organized as a multiple-inheritance hierarchical graph of concepts—OBJECTS and EVENTS—each of which is described using PROPERTYs. Concepts are named using language-independent labels, written in small caps, that resemble English words only for the benefit of English-speaking developers. The actual meaning of a concept is its inventory of property-facet-filler triples.

Facets permit the ontology to include an extra level of detail about property fillers, such as the fact that the most typical colors of a car are white, black, silver, and gray; other normal, but less common, colors are red, blue, brown, and yellow; and rare colors are green and purple. The inventory of facets includes: *default*, which represents the most restricted, highly typical subset of fillers; *sem*, which represents typical selectional restrictions; *relaxable-to*, which represents what is, in principle, possible although not typical; and *value*, which represents not a constraint but an actual, non-overridable value. *Value* is used primarily in episodic memory but applies to a select few properties in the ontology, such as DEFINITION, IS-A, and SUB-CLASSES. A sampling of properties from the ontological frame for the event SURGERY illustrates the use of facets.

SURGERY

DEFINITION	value	performing a medical procedure that involves cutting into tissue
IS-A	value	INVASIVE-PROCEDURE
AGENT	default	SURGEON
	sem	MEDICAL-PERSONNEL
	relaxable-to	HUMAN
THEME	default	HUMAN
	sem	ANIMAL

INSTRUMENT	sem	SURGICAL-INSTRUMENT
LOCATION	default	OPERATING-ROOM
	sem	MEDICAL-BUILDING
	relaxable-to	PLACE

The main benefits of writing an ontology in a knowledge representation language rather than a natural language are (a) the absence of ambiguity in the knowledge representation language, which makes the knowledge suitable for automatic reasoning, and (b) the ontology's reusability across natural languages.

The ontology covers both general and specific domains and currently contains around nine thousand concepts. The bulk of it was compiled some twenty-five years ago, and, in keeping with our lab's focus on cognitive modeling research rather than application development, only modest enhancements have been made since. We use the ontology as a research tool and enhance it to test the ever-growing inventory of microtheories—including the hypothesis that LEIAs can acquire ontology independently through dialog, experience, and reading.

The example of SURGERY illustrates the simplest kind of ontological structure. The expressive power of the ontology is actually far greater, and its content far richer, since the ontology is intended to support multiple agent functionalities including simulation, reasoning, learning, teaching, explaining, and beyond.

3.2.1 Properties

Every OBJECT and EVENT is described using PROPERTYS. The inventory of PROPERTYS essentially supplies the axiomatic layer of the ontology's representational system. PROPERTYS are characterized by their DOMAIN (constraints on the sets of concepts for which they are defined) and their RANGE (their value sets). The LEIA ontology includes several types of PROPERTYS:

- IS-A and SUBCLASSES indicate the concept's placement in the inheritance hierarchy. Multiple inheritance is permitted but not overused, and rarely does a concept have more than two parents.

- RELATIONS indicate relationships among OBJECTS and EVENTS. The DOMAIN and RANGE of RELATIONS are, therefore, filled by OBJECTS and/or EVENTS. Examples include the case roles (e.g, AGENT, THEME[7]), spatial relations (e.g., ABOVE, NEXT-TO), HAS-OBJECT-AS-PART, CAUSED-BY, and so on. All RELATIONS have inverses: for example, the inverse of AGENT is AGENT-OF, the inverse of ABOVE-AND-TOUCHING is BELOW-AND-TOUCHING, and the inverse of CAUSED-BY is EFFECT.

- SCALAR-ATTRIBUTEs are properties of OBJECTs and EVENTs that can be expressed by numbers or ranges of numbers, such as COST, VELOCITY, and WEIGHT. Values can be actual—for example, *180 pounds* is "WEIGHT (180 MEASURED-IN POUND)"—or they can be expressed on the abstract scale {0,1}—for example, *heavy* is "WEIGHT .8" and *light* is "WEIGHT .2."

- LITERAL-ATTRIBUTEs are properties of OBJECTs and EVENTs whose fillers are represented by uninterpreted literals. For example, the property MARITAL-STATUS has the literal fillers *single, married, divorced*, and *widowed*.

- SUBEVENTS holds ontological scripts, also known as complex events (section 3.2.4).

- SEMANTIC-EXPANSION holds concept-based descriptions of complex properties (section 3.2.1).

- The DEFINITION field holds a natural language string that explains the concept. Definitions are used by developers as well as by LEIAs for purposes of explanation.

Ontological PROPERTYs can also function as ABSTRACT-OBJECTs: *Friendliness is important; Color livens up a house.* Property-based abstract nouns are recorded in the lexicon as an ABSTRACT-OBJECT with a RELATION to the meaning of the adjective. Since the adjective *friendly* is described as "FRIENDLINESS .8," the noun *friendliness* is described as:

ABSTRACT-OBJECT
   RELATION   FRIENDLINESS-1
FRIENDLINESS-1
   RANGE   .8

Treating nominal uses of properties as ABSTRACT-OBJECTs allows them to participate in larger propositions in the normal way—they can be modified, evaluated as case-role fillers, and so on. For example, the meaning representation for **Friendliness is important** is:

ABSTRACT-OBJECT-1
   RELATION   FRIENDLINESS-1
   IMPORTANCE   .8
FRIENDLINESS-1
   RANGE   .8

Properties can be simple or complex. Simple properties, like WEIGHT and VELOCITY, can be directly grounded in the real world, without further ontological decomposition. Complex properties, by contrast, can be explained in terms of other OBJECTS, EVENTS, and PROPERTYS. For example, a salient feature of people is who they are married to, which is expressed by the RELATION called HAS-SPOUSE. This is not a primitive; it can be explained as the state that is the EFFECT of a MARRY event. If you know about the MARRY event, you can infer the HAS-SPOUSE relation and vice versa.

Complex properties are essential for cognitive modeling because they capture how people think about the world—and, in turn, how they talk, teach, learn, and reason about it. Recording the semantic interpretations of complex properties in the ontology enables agents to make inferences and understand implicatures like people do—an essential capability that has largely eluded AI to date.[8] For example, if an agent hears *Jan and Paul are married*, which is semantically analyzed using the HAS-SPOUSE relation, it must also understand that they were both AGENTS of the same MARRY event. Similarly, if an agent hears that something is large, which is semantically analyzed as "SIZE .8," it must understand (a) that this implies a high value of one or more of the primitive properties LENGTH, WIDTH, DEPTH, HEIGHT, and/or WEIGHT, and (b) what the ballpark size of the given object is, since a large beetle is much smaller than a large oak tree.

Philosophical analysis of the nature and classification of properties is beyond the scope of book.[9] Here, we focus on how LEIAs are being prepared to interpret complex properties and make property-related inferences so that they can communicate, reason, and learn about the world like people do.

For orientation, the property-related phenomena we will describe are as follows:

− Complex properties can be states resulting from an event.
− Complex properties can generalize over repeated events.
− Complex properties can be shortcuts for an event-based chain.
− Complex properties can generalize over other properties.
− Abstract values of scalar properties can be calculated.
− Complex properties can have indirect semantic expansions.
− Qualitative properties can be used with quantitative implications.

Complex Properties Can Be States Resulting from an Event   Typical examples of states that result from events involve familial relations. Consider again the example of getting married. Below is an excerpt from the ontological description of the event MARRY that shows its relation to the state HAS-SPOUSE. The numerical indices represent ontological instances, which allow for coreference and disambiguation within multi-frame ontological structures (see section 3.2.2).

```
MARRY
    AGENT           HUMAN-1, HUMAN-2
    EFFECT          HAS-SPOUSE-1, HAS-SPOUSE-2
HAS-SPOUSE-1
    DOMAIN          HUMAN-1
    RANGE           HUMAN-2
HAS-SPOUSE-2
    DOMAIN          HUMAN-2
    RANGE           HUMAN-1
```

This says that when two people are the AGENTs of MARRY, the EFFECT is that they are in a HAS-SPOUSE relationship. The connection between MARRY and HAS-SPOUSE is also recorded in the SEMANTIC-EXPANSION field of HAS-SPOUSE's property definition, as shown below. Explanatory comments are provided after semicolons.

```
HAS-SPOUSE
    DOMAIN          HUMAN-1
    RANGE           HUMAN-2
    INVERSE         SPOUSE-OF
    SEMANTIC-EXPANSION
        PRECONDITION   MARRY-1, MODALITY-1   ; they were married and not divorced
        MARRY-1                              ; they were married
          AGENT        HUMAN-1, HUMAN-2
        MODALITY-1                           ; there was no divorce
          TYPE         EPISTEMIC
          VALUE        0
          SCOPE        DIVORCE-1
        DIVORCE-1
          AGENT        HUMAN-1, HUMAN-2      ; by these same people
          TIME         > MARRY-1.TIME        ; since their marriage
```

The SEMANTIC-EXPANSION of HAS-SPOUSE includes more information than the MARRY script—namely, that the individuals were not subsequently divorced.

To recap: the reason why the ontology includes the relationship HAS-SPOUSE is that people think and talk about the world in terms of kinship relations. The relationship between HAS-SPOUSE and MARRY must be explicitly recorded in the ontology to support the bidirectional inferencing between getting married, which is an EVENT, and being married, which is a state expressed as a RELATION.

Complex Properties Can Generalize over Repeated Events   Athletes have coaches, people have dentists, and kids have babysitters. These social relations—HAS-COACH, HAS-DENTIST, HAS-BABYSITTER—imply repeating events and, in some cases, a formal process of establishing the relationship, such as filling out paperwork to become a dentist's patient. Seeing a dentist for an emergency treatment while traveling does not make that person *one's dentist*.

The correlation between such properties and the event sequences that they imply can be expressed in two ways in the ontology. On the one hand, it can be appended to the event description as a conditional statement. If we assume that more than one instance of coaching is needed to infer a HAS-COACH relationship (the actual number can be understood differently by different people), then it will look as follows:

```
COACHING-EVENT-1
    AGENT             COACH-1
    BENEFICIARY       ATHLETE-1
    SEMANTIC-EXPANSION
      If                                    ; If
        SET                                 ; there is more than one coaching event
          MEMBER-TYPE    COACHING-EVENT-1
          CARDINALITY    >1
          COACHING-EVENT-1                  ; in which
            AGENT        COACH-1            ; a particular coach coaches
            BENEFICIARY  ATHLETE-1          ; a particular athlete
      Then                                  ; Then
        HAS-COACH-1                         ; that athlete has that coach
          DOMAIN         ATHLETE-1
          RANGE          COACH-1
```

This information is also stored in the SEMANTIC-EXPANSION field of the description of HAS-COACH, which also lists another way of establishing the relationship: by hiring the coach.

```
HAS-COACH-1                              ; A particular athlete has a particular coach
   DOMAIN    ATHLETE-1
   RANGE     COACH-1
   INVERSE   COACH-OF
   SEMANTIC-EXPANSION
      If                                 ; If
         Either                          ; the coach has coached the athlete more than once
            SET
               MEMBER-TYPE   COACHING-EVENT-1
               CARDINALITY   >1
            COACHING-EVENT-1
               AGENT            COACH-1
               BENEFICIARY      ATHLETE-1
         Or                              ; or
            HIRE-COACH                   ; the athlete hired the coach
               AGENT    ATHLETE-1
               THEME    COACH-1
      Then                               ; Then
         HAS-COACH-1                     ; the athlete has the coach
            DOMAIN    ATHLETE-1
            RANGE     COACH-1
```

For an example that does not involve social relations, we can look at the Maryland Virtual Patient clinician training system (section 8.5), where patient symptoms were modeled as properties whose values changed throughout the interactive simulation. For example, DIFFICULTY-SWALLOWING had the patient as its DOMAIN and the abstract values {0,1} as its RANGE. At the beginning of a simulation run, before the patient experienced any symptoms, the value for DIFFICULTY-SWALLOWING was 0; but if the patient had a disease that caused difficulty swallowing, as the disease progressed, the value for this property would increase. This property captured how physicians think, talk, and reason about this symptom. In a word, they generalize— they do not think in terms of the innumerable times a patient swallows in a given day, month, or year. However, in order for the property DIFFICULTY-SWALLOWING to have meaning, it must be described in the ontology with reference to the implied large set of SWALLOW events:

```
DIFFICULTY-SWALLOWING-1
   DOMAIN    HUMAN-1                     ; A particular person has
   RANGE     {0,1}                       ; a particular value for DIFFICULTY-SWALLOWING
   SEMANTIC-EXPANSION
      If                                 ; If
```

DIFFICULTY-ATTRIBUTE-1		; difficulty-attribute applies
DOMAIN	SET-1	; to a set of swallow events
RANGE	**var1**	; and has a particular value
SET-1		; and that set
MEMBER-TYPE	SWALLOW-1	; of swallow events
QUANT	.8	; is large
SWALLOW-1		
AGENT	HUMAN-1	; and is carried out by this person
Then		; Then
DIFFICULTY-SWALLOWING-1		; the value of difficulty-swallowing is the
DOMAIN	HUMAN-1	; range of difficulty-attribute-1—that is,
RANGE	**var1**	; how difficult the swallowing is

This says, "If the value of DIFFICULTY-ATTRIBUTE for a large number of SWALLOW events by a particular person is [some value], then the value of DIFFICULTY-SWALLOWING for that person is [that same value]."

Complex Properties Can Be Shortcuts for an Event-Based Chain   Continuing to draw examples from the medical domain, when clinicians think, talk, and teach about DISEASES, properties like the following are useful: HAS-TYPICAL-SYMPTOM, HAS-DIAGNOSTIC-TEST, SUFFICIENT-GROUNDS-TO-SUSPECT, SUFFICIENT-GROUNDS-TO-DIAGNOSE, SUFFICIENT-GROUNDS-TO-TREAT, and PREFERRED-ACTION-WHEN-DIAGNOSED. These, in fact, were included in the Maryland Virtual Patient system mentioned earlier, and they are conceptual shortcuts for what is actually going on. Whereas HAS-TYPICAL-SYMPTOM links a disease to a symptom, in reality diseases are not directly associated with symptoms: PATIENTS who are experiencing a DISEASE also likely experience the given SYMPTOM (both DISEASES and SYMPTOMS are EVENTS). This explanation is recorded in the SEMANTIC-EXPANSION zone of the property called HAS-TYPICAL-SYMPTOM:

HAS-TYPICAL-SYMPTOM		
DOMAIN	DISEASE-1	
RANGE	SYMPTOM-1	
SEMANTIC-EXPANSION		
If		
ANIMAL-1		
EXPERIENCER-OF	DISEASE-1	
Then		
ANIMAL-1		
EXPERIENCER-OF	SYMPTOM-1	
SYMPTOM-1		
LIKELIHOOD	.8	

Recording the likelihood of the symptom as .8 (on the abstract scale {0,1}) conveys the notion of *typical*. For each particular disease, each particular symptom has a population-based likelihood, which will be recorded if the agent has learned that information.

Complex Properties Can Generalize over Other Properties   When thinking about families, one thinks about grandparents, great-grandparents, cousins, aunts, siblings, and the rest. However, all of these can be explained in terms of the more primitive properties HAS-OFFPSRING and HAS-SPOUSE. The property HAS-GREAT-GRANDCHILD illustrates how properties can, and need to be, explained in terms of more primitive properties.

HAS-GREAT-GRANDCHILD
   DOMAIN   HUMAN
   RANGE   HUMAN
   INVERSE   GREAT-GRANDCHILD-OF
   SEMANTIC-EXPANSION
    If HUMAN-1 (HAS-GREAT-GRANDCHILD HUMAN-4)
    Then HUMAN-1 (HAS-OFFSPRING HUMAN-2)
     And HUMAN-2 (HAS-OFFSPRING HUMAN-3)
     And HUMAN-3 (HAS-OFFSPRING HUMAN-4)

In addition, since HAS-OFFSPRING is, itself, semantically expanded using the events BEAR-OFFSPRING and MARRY, the agent has the knowledge to reason about these events if needed.

Abstract Values of Scalar Properties Can Be Calculated   It is natural to think and talk about the world in an underspecified way: *a tall person, a fast race, an inexpensive meal*. Abstract values of scalar attributes allow us to represent correspondingly imprecise meanings: *a tall person*—"HUMAN (HEIGHT .8)," *a fast race*—"RACE (VELOCITY .8)," *a moderately-priced meal*—"MEAL (COST .5)." However, in some cases, underspecified descriptions need to be concretized in order to serve the needs of agent reasoning. For example, if we ask a robotic LEIA to dig a hole the size of a large packing box, it needs to convert that description into some actual LENGTH, WIDTH, and DEPTH. For the agent to make such calculations, it must know the actual size of the object referred to. For example, packing boxes sold by one US company[10] range from $6 \times 6 \times 6$ inches to $24 \times 24 \times 24$ inches, which informs the following ontological description:

PACKING-BOX
   LENGTH   sem 6–24 (MEASURED-IN INCH)
   WIDTH   sem 6–24 (MEASURED-IN INCH)
   HEIGHT   sem 6–24 (MEASURED-IN INCH)

Calculations of actual sizes from relative ones, which are always understood to be approximate, are straightforward:[11]

A large box (abstract value .8) is around 20.4 inches L, W, H

A small box (abstract value .2) is around 9.6 inches L, W, H

The function for calculating actual values from abstract ones is recorded as a SEMANTIC-EXPANSION attached to the SIZE property, which is the source of such generalizations. We can see that this works pretty well in a completely different domain—people's heights. If we say that the typical range of heights for people is 4'10" (58") to 6'4" (76"), then a tall person (.8 on the scale) is around 6' and a short person is around 5'1".

Although generic formulas are useful, they do not work well in all cases, as when extreme high and/or low values are substantially distant from the normal range.[12] Consider cars: a Rolls-Royce Boat Tail car costs $28 million, whereas the cheapest new Kia is under $17,000, and you can buy an old junker for a couple hundred bucks.[13] The ontology permits recording such values explicitly using facets: *sem* for the normal values and *relaxable-to* for the extreme ones, which can improve the relative calculations by orienting around the *sem* range of values. But if agents need to be able to reason more precisely about such values—in a way similar to people—then it is also possible to explicitly list understood values as semantic expansions of the given property.

Complex Properties Can Have Indirect Semantic Expansions   A repeating theme in this section is that the ontology should record how people think about the world, which is clear from how they talk about it. Imagine that a doctor is teaching students about the disease achalasia and says, "Patients with achalasia complain of difficulty swallowing." What the students will glean from this is that a typical symptom of achalasia is difficulty swallowing. In ontological terms, this is:

ACHALASIA
    HAS-TYPICAL-SYMPTOM   DYSPHAGIA

But how do the students extract the intended meaning from what is actually said?

- They know that the linguistic construction "Patients with $NP_{DISEASE}$ complain of $NP_{SYMPTOM}$" means that patients *report* a symptom, not that they whine about it. (The subscripts in the presentation of constructions indicate ontological constraints on the meaning of the constituents.)

- They know how to reason about generics: if patients with a given disease report a symptom, then the symptom is typical of the disease.

The question is, how best to prepare an agent to do this reasoning? The fastest and most reliable way is to record typical ways of thinking and talking about property values as formal representations in the SEMANTIC-EXPANSION zone of the PROPERTY. Typical ways of expressing HAS-TYPICAL-SYMPTOM include the following, among others.

- Patients with $NP_{DISEASE}$ complain of $NP_{SYMPTOM}$
- $NP_{SYMPTOM}$ suggests $NP_{DISEASE}$
- $NP_{SYMPTOM}$ is suggestive of $NP_{DISEASE}$
- $NP_{SYMPTOM}$ is key to diagnosing $NP_{DISEASE}$

The SEMANTIC-EXPANSION zone of HAS-TYPICAL-SYMPTOM lists the formal meaning representations of such formulations, which are recorded in the agent's lexicon as well so that agents can link language inputs to ontological knowledge. Taking the first one as an example:

HAS-TYPICAL-SYMPTOM
  SEMANTIC-EXPANSION

If		; If
DECLARATIVE-SPEECH-ACT-1		
AGENT	PATIENT-1	; patients say
THEME	SYMPTOM-1	; that there is a symptom
SYMPTOM-1		
EXPERIENCER	PATIENT-1	; that they are experiencing
PATIENT-1		; and the same patients
CARDINALITY	$>1^{14}$	; (indicates plurality)
EXPERIENCER-OF	DISEASE-1	; have a particular disease
Then		; Then
DISEASE-1		; the disease
HAS-TYPICAL-SYMPTOM	SYMPTOM-1	; has this as a typical symptom

We record semantic expansions like this because, at the current state of the art, there is no other way for agents to predict all of the different ways that people think and talk about properties, and recording them is a time-efficient way of making systems work reliably in the near term. For example, if we want a LEIA to learn about a large number of diseases by reading texts and listening to teaching physicians, it makes sense to do the small amount of knowledge acquisition that prepares for common eventualities like the ones above because it promises a good payoff across hundreds of diseases.

Returning to the issue of hybridization, data-driven tools could be used to collect examples for the agent to learn from by identifying and clustering excerpts that include symptoms and diseases. Then the LEIA could semantically analyze them into text meaning representations, cluster those meaning representations based on the concepts they use, and present the results to a knowledge engineer to vet as candidate values for the SEMANTIC-EXPANSION zone of HAS-TYPICAL-SYMPTOM. And so on, for other properties of interest.

Qualitative Properties Can Be Used with Quantitative Implications   Qualitative properties such as SPATIAL-RELATIONS (e.g., ABOVE, BELOW, ADJACENT-TO) can carry quantitative implications that an agent needs to understand.[15] For example, the property NEAR is a relation that compares the locations of two physical objects. Although it does not assert any particular distance between them, the implied distance depends on the sizes of the objects in question. In the following examples, the distances implied would be best measured in inches, feet, small numbers of miles, and tens of miles, respectively.

(3.1) The pencil is near the notebook.
(3.2) The car is near the stop sign.
(3.3) Her house is near her high school.
(3.4) My hometown is near yours.

It can be important for agents to understand such calculations for similar reasons as for our hole-digging robotic LEIA. If a robotic LEIA is told to stand near the door, does that mean a couple of inches away or a dozen yards away? The way to prepare the robot to make this calculation is to formulate the semantics of nearness.[16] The following is a first approximation:

For each of the OBJECTS being compared
   Take whichever values of LENGTH, WIDTH, DEPTH, HEIGHT are known.[17]
   Average them together.
   NEAR is <= 1.5 * average.

This calculation is a model—a simplistic one, to be sure—that produces reasonable results for three of the four examples above: A pencil near a notebook is <= 13.8 inches away; A car near a stop sign is <= 13.8 feet away; and a hometown near another hometown is <= 17.5 miles away.[18] For the house near the high school example, however, this formula does not work: a house that is near a high school is much farther away than 1.5 times their average size, even if one counts their grounds. People have, and agents need, additional knowledge about what it means for a building to be located near another building. Moreover, there are subclasses: buildings that are near each other on a college campus are likely to be closer to each other than homes that are near each other in suburbia. The importance of reasoning about actual distances is illustrated using the example of a LEIA learning rules of the road in section 7.1.5.

Recapping Why Properties Are So Important—And No Simple Matter   The expressive power of the property apparatus in the LEIA ontology is key to modeling agents that learn, reason, and communicate like people. It allows for the formalization of an interesting analogy between complex properties and habitual or reflexive actions. When an action is habitual or reflexive, people don't think about it anymore—unless something about the situation triggers special attention. Thus, we take the same route home from work every day unless a traffic jam causes us to replan, and we drive a stick shift car automatically until somebody asks us to teach them how to do it. Similarly, we use complex properties as shortcuts for learning and reasoning but can explain them if need be. For example, we think about our grandmother without thinking about a sequence of birthing and marriage events, but we could explain the relationship in terms of those events if asked to. Since cognitive modeling involves hypothesizing about what people seem to know and how they seem to reason, there is ample justification for including in the ontology whichever properties people orient around and explaining them in ways that prepare agents to reason about them in same ways as people do.

3.2.2   Ontological Instances

Ontological instances are remembered occurrences of ontological concepts that are used for coreference and reification within larger ontological descriptions. (Reification is filling a property's slot with a complex structure.) For example, whereas it is correct to say that CARS have TIRE as their part, it is more informative to specify that there are four of them.

CAR
   HAS-OBJECT-AS-PART   sem   TIRE-1
TIRE-1
   CARDINALITY   value   4

Another example of the use of ontological instances involves typical sequences of events. For example, after someone asks a yes-no question, the other person typically answers it. This question-answer combination is an example of an adjacency pair—a topic that will be further discussed in chapter 6.

REQUEST-INFO-YN

AGENT	sem	HUMAN
	coref	RESPOND-TO-REQUEST-INFO-YN-1.BENEFICIARY
BENEFICIARY	sem	HUMAN
	coref	RESPOND-TO-REQUEST-INFO-YN-1.AGENT
ADJACENCY-PAIR	default	RESPOND-TO-REQUEST-INFO-YN
	coref	RESPOND-TO-REQUEST-INFO-YN-1

RESPOND-TO-REQUEST-INFO-YN

AGENT	sem	HUMAN
	coref	REQUEST-INFO-YN-1.BENEFICIARY
BENEFICIARY	sem	HUMAN
	coref	REQUEST-INFO-YN-1.AGENT
ADJACENCY-PAIR-OF	default	REQUEST-INFO-YN
	coref	REQUEST-INFO-YN-1

We call semantically linked pairs of events like the one above *scriptlets*—small, script-like structures—and they have some noteworthy features. First, since people know such pairs of events, so, too, must LEIAs: *X asks Y a question* → *Y answers it*; *X tells Y a joke* → *Y laughs*; *X waves to Y* → *Y waves back*. Second, these event sequences are domain independent: no matter when or where somebody asks a question, the likely next move is for the other person to answer it. Third, from the point of view of knowledge engineering, it is useful to model typical sequences of events that are reusable across domains and applications.

3.2.3 Proto-Instances

Proto-instances are a hybrid between an ontological concept and a concept instance. Like concepts, they are generic and are a proper part of the ontology. Like instances, they are more specific than basic ontological descriptions in that certain property values are asserted. The need for proto-instances becomes clear when we consider some of the applications that agents can participate in. For example, simulation-based training systems offer trainees a large variety of practice cases that differ with respect to salient feature values. Each case thus defined is a proto-instance that can be instantiated time and time again by different trainees in different simulation runs. As a concrete example, our Maryland Virtual Patient clinician training application featured an inventory of virtual patients, each of which was a proto-instance. In training systems, proto-instances allow teachers to encapsulate different teaching scenarios, such as a GERD patient who will progress to adenocarcinoma if left untreated.

3.2.4 Scripts

Ontological scripts record complex events along with their participants and props.[19] They can reflect knowledge in any domain—what happens at a doctor's appointment, how to build a chair, what to do at a four-way stop; and they can be at any level of specificity—from a basic sequence of events to the level of detail needed to generate a computer simulation. Their descriptions include more expressive means than the simple frame illustrated by SURGERY above. For example, scripts can require the coreferencing of arguments across events, they can have optional and variously ordered events, they can require time management, and so on. Scripts can be recorded by knowledge engineers, or they can be acquired by agents on the fly.[20]

Scripts both guide agent operation and support their reasoning about the world. Example (3.5) illustrates how script-based knowledge is needed for making implicatures during language understanding.

(3.5) "How was your doctor's appointment?" "Great! The scale was broken!"

Why does the second speaker say *the scale*? What licenses the use of *the*, considering that this object was not previously introduced into the discourse? The mention of a doctor's appointment prepares the listener to mentally access objects, like scale, and events, like getting weighed, that are typically associated with a doctor's appointment, making those objects and events primed for inclusion in the situation model. In fact, the linguistic licensing of *the* with *scale* is evidence that such script activation actually takes place. Of course, it is script-based knowledge that also explains why the person is happy—and it further allows us to infer the body type of the speaker. If we want LEIAs to be able to reason at this level as well, then scripts are the place to store the associated knowledge.

When the knowledge in a script is used to guide agent action, the agent must create an instance of it, which is called a plan. A plan differs from a script in that (a) it selects a particular path through the often-variable sequence of events permitted by the script and (b) it fills the events' case roles in particular ways.

The events and objects referred to in a script are, themselves, concepts that are recorded in appropriate branches of the ontology. This means that scripts are organized like well-constructed computer programs—not as a massive main function but, rather, as hierarchically organized drilldowns of scripts that ultimately end in singleton events or function calls.

We will describe scripts using the example of the AGENT-FUNCTIONING-FLOW script that implements the agent's cognitive architecture and makes the agent self-aware so that it can explain its own functioning (cf. section 2.1). This script uses the EVENTS and OBJECTS shown in the ontological subtrees in table 3.2, which we provide as a crib to be consulted when reading the scripts themselves. This description of AGENT-FUNCTIONING-FLOW serves the double duty of presenting an example of a script as an ontological entity and describing an important aspect of LEIA operation.

The top level of the AGENT-FUNCTIONING-FLOW script has five ordered subevents that correlate with the architecture diagram in figure 2.1.[21]

AGENT-FUNCTIONING-FLOW
  DEFINITION This script implements the LEIA's cognitive architecture.
  AGENT   LEIA-1
  SUBEVENTS
      1. PERCEPTION-RECOGNITION
      2. PERCEPTION-INTERPRETATION
      3. DELIBERATION
      4. ACTION-SPECIFICATION
      5. ACTION-RENDERING

All of the subevents are, themselves, scripts, which we will describe in turn. The first one is PERCEPTION-RECOGNITION, whose function is described in its definition field.

PERCEPTION-RECOGNITION
  DEFINITION The agent determines which recognizer is needed to process a signal, runs it, and stores the resulting data.
  AGENT   LEIA
  INPUT   PERCEPTION-INPUT
  OUTPUT   PERCEIVED-DATA

SUBEVENTS
 TRY: recognize-text-input
  EXPL "text input is recognized and stored as data"
 TRY: recognize-speech
  EXPL "speech input is recognized and stored as data"
 TRY: recognize-visual-input
  EXPL "visual input is recognized and stored as data"
 TRY: recognize-interoception
  EXPL "interoceptive input is recognized and stored as data"

The input to PERCEPTION-RECOGNITION is any kind of PERCEPTION-INPUT which, as indicated by that concept's subclasses in table 3.2, currently includes

Table 3.2
The ontological subtrees of EVENTS and OBJECTS that are used in the AGENT-FUNCTIONING-FLOW script.

EVENTS	OBJECTS	
AGENT-FUNCTIONING	AGENT-SPECIFIC-OBJECTS	
AGENT-FUNCTIONING-FLOW	AGENT-COGNITION-TOOL	
AGENT-PERCEPTION-EVENT	INTEROCEPTION-PROCESSOR	
PERCEPTION-RECOGNITION	NLU-SYSTEM	
PERCEPTION-INTERPRETATION	VISION-PROCESSOR	
INTEROCEPTION	PERCEPTION-INPUT	
NATURAL-LANGUAGE-UNDERSTANDING	INTEROCEPTION-SIGNAL	
VISION-INTERPRETATION	SPEECH-SIGNAL	
AGENT-REASONING-EVENT	TEXT-INPUT	
AGENT-PLANNING	VISION-INPUT	
MEMORY-MANAGEMENT	PERCEIVED-DATA	
DELIBERATION	INTEROCEPTION-DATA	
PROCESS-DAEMONS	SPEECH-DATA	
AGENT-ACTION-EVENT	TEXT-DATA	
ACTION-SPECIFICATION	VISION-DATA	
CONVERT-MMR-TO-GMR	XMR	; meaning representation
CONVERT-MMR-TO-AMR	AMR	; robotic action MR
CONVERT-MMR-TO-SMR	GMR	; generation MR
RENDERING	IMR	; interoception MR
GENERATE-GMR	MMR	; mental MR
LAUNCH-ROBOTIC-EFFECTOR	SMR	; simulated action MR
SIMULATE-PHYSICAL-ACTION	TMR	; text MR
	VMR	; vision MR

interoception, speech, text, and vision. The output is the associated kind of PERCEIVED-DATA, whose subclasses are also shown in the table.

There are four subevents, which are actually conditions with different preconditions that are evaluated in turn; this is the semantics of "TRY." The SUBEVENTS field says, "If this is text input, then do text-input recognition; Else if this is speech input, then do speech-input recognition," and so on. These subevents are not concepts, they are pointers to code that carries out the associated functions. Their status as pointers to code is indicated by the lowercase font. Since they are not concepts, the agent cannot look up their definitions to know what they mean and what they do. So, to enable the agent to explain these actions, a metadata field called "EXPL" (explanation) provides a short description.

Using concepts vs. procedure calls as subevents of scripts

A subevent of a script is recorded as a concept if:

a. it is, itself, a script; or

b. it is a non-decomposable event that has a freestanding status in the ontology.

By contrast, a subevent of a script is recorded as a procedure call if:

a. it implements a procedure that is below the threshold of what the agent needs to understand;

b. it implements a procedure that is not explainable because it is grounded in machine learning; or

c. it implements a procedure that should eventually be described using a concept but is temporarily being treated as an opaque function in order to speed up the implementation of a particular system.

It is important, methodologically, not to attempt to make every line of code needed to implement LEIAs fully understood and explainable by them. This would be an inefficient use of resources. Instead, agents should be self-aware to a useful degree, and they should be prepared to explain their behavior in useful ways.

When the agent instantiates PERCEPTION-RECOGNITION as a plan, that plan reflects a specific path through the script, depending on which type of input was recognized. If a language input was recognized, the agent launches NATURAL-LANGUAGE-UNDERSTANDING; if a visual input was recognized, the agent launches VISION-INTERPRETATION; if a bodily sensation was recognized, the agent launches INTEROCEPTION; and so on for other perception modalities that could be implemented.

PERCEPTION-INTERPRETATION

DEFINITION	The agent analyzes data into the appropriate type of XMR.	
AGENT	LEIA	
INPUT	PERCEIVED-DATA *from* PERCEPTION-RECOGNITION.OUTPUT	
OUTPUT	XMR	
INSTRUMENT	AGENT-COGNITION-TOOL	
SUBEVENTS		
TRY:	NATURAL-LANGUAGE-UNDERSTANDING	
TRY:	VISION-INTERPRETATION	
TRY:	INTEROCEPTION	

The OUTPUT of PERCEPTION-INTERPRETATION is some type of meaning representation, an XMR, but the actual type depends on the channel of perception: it might be a TMR, a VMR, an IMR, and so on. In describing the rest of the script, we will not continue to highlight the distinction between the static script descriptions that populate the ontology and the plans that an agent dynamically generates, but this distinction should be kept in mind.

Once the agent has understood an input, it needs to decide what to do in response to it. That is handled by the DELIBERATION script, which takes the just-generated XMR (e.g., TMR, VMR) as input and outputs a mental meaning representation (MMR) that records its decision about what to do.

DELIBERATION

DEFINITION	The agent decides how to respond to an input.
AGENT	LEIA
INPUT	XMR *from* PERCEPTION-INTERPRETATION.OUTPUT
OUTPUT	MMR
SUBEVENTS	

TRY: run-procedure-from-concept-in-XMR

    EXPL "The XMR contains a concept that, when instantiated, triggers a particular response."

    EX "If someone yells 'Fire!,' the triggered response is to exit the building."

TRY: act-on-adjacency-pair

    EXPL "The XMR contains a concept that has an adjacency pair, which indicates the default response type."

    EX "If the TMR includes REQUEST-INFO-WH, the adjacency pair is RESPOND-TO-REQUEST-INFO-WH; that is, the default response to a wh-question is to answer it."

TRY: PROCESS-DAEMONS
TRY: continue-plan-on-agenda
  EXPL   "The plan currently on agenda is continued."

The first condition checks to see if the XMR contains any concepts whose ontological descriptions indicate a necessary event in response. For example, if a TMR includes the speech act WARN-OF-FIRE—which will be generated, for example, by someone yelling "Fire!"—then this triggers the agent to exit the building.

The second condition exploits the adjacency pairs recorded in the ontology. As mentioned earlier, adjacency pairs reflect typical sequences of events that serve as an agent's default response. For example, if X asks a question, Y answers it; if X holds out his hand, Y shakes it; and so on. Adjacency pairs drive dialog interactions, as illustrated in chapters 6–8.

The third condition is a script that checks if any of the agent's daemons are triggered by the XMR. A daemon is a procedure that is on agenda and is available to be run any time its preconditions are fulfilled. If a daemon is triggered, the agent decides what to do in response. For example, say the agent is asked to decide whether to agree to a medical procedure and it has a daemon on its agenda that requires it to know about the associated pain of procedures before agreeing to them. If the agent does not know how painful the procedure is, then it must find that out before making a decision. This bit of processing is formulated as a script (PROCESS-DAEMONS) rather than a procedure call because it has its own subevents and decision functions.

The last condition covers the situation in which the latest input does not require action. In this case, the agent continues to pursue whatever goal it was pursuing prior to the last perceptual input.

DELIBERATION results in a mental meaning representation (MMR) that contains the agent's decision about *what* to do next but not yet *how*. For example, if it was asked a yes-no question, the MMR might represent its intention to convey a negative response, but there are various ways it can do that, as by speech, text, or body language. And, for each of these modalities, there are subsequent decisions to be made. For example, if the agent chooses speech, how polite will the response be, and will the agent provide the reason for it? If the agent chooses body language, then which gesture will it use, and how emphatically will it enact that gesture? All of this is decided in the script called ACTION-SPECIFICATION, whose subevents refer to the specific types of target representations that can be generated: GMRs for language generation, AMRs for robotic action, or SMRs for simulated physical action.

ACTION-SPECIFICATION

DEFINITION	The agent decides which type of action to use to convey the MMR and carries out the reasoning to convert the MMR into the associated type of XMR.
EXAMPLE	If the agent wants to respond negatively to a yes/no question, it has to decide whether to use language, body language, or both; and once it has decided, it has to record that meaning/intention in a generation meaning representation (GMR), an action meaning representation (AMR), or a simulated action meaning representation (SMR).
AGENT	LEIA
INPUT	MMR *from* DELIBERATION.OUTPUT
OUTPUT	XMR
SUBEVENTS	

TRY:	CONVERT-MMR-TO-GMR	; for language generation
TRY:	CONVERT-MMR-TO-AMR	; for robotic action
TRY:	CONVERT-MMR-TO-SMR	; for simulated action

The final stage of AGENT-FUNCTIONING-FLOW is ACTION-RENDERING, in which the agent actually generates text, speech, physical action, or simulated action.

ACTION-RENDERING

DEFINITION	The agent carries out the action represented in the action-oriented XMR.
EXAMPLE	If the agent decides to respond positively to a yes/no question using language, it has to create a sentence to reflect the meaning of the GMR.
AGENT	LEIA
INPUT	XMR *from* ACTION-SPECIFICATION.OUTPUT
OUTPUT	EVENT
SUBEVENTS	

TRY:	LANGUAGE-GENERATION
TRY:	ROBOTIC-ACTION
TRY:	SIMULATED-ACTION

When the action involves language, ACTION-RENDERING includes creating the actual sentence that will realize the meaning that was fully specified in the GMR (see section 4.3 for details). When the action involves robotic or simulated action, ACTION-RENDERING involves creating the signals to pass to the associated effectors.

Let us recap some important points about scripts.

- Scripts can be recorded by knowledge engineers, or they can be learned by agents during their operation.
- Scripts neatly open up into subscripts, which organizes knowledge representation.
- Scripts are as concept-based and explainable as possible but as stream-lined and program-based as necessary. It would be a poor use of time to attempt to make every line of code that implements agent action fully explainable by the agent.
- When a script is instantiated as a plan, it reflects a specific path through the script and requires such things as handling coreferences of partici-pants and props across multiple events.
- Like all of the agent's knowledge, scripts are fully inspectable and modi-fiable over time.
- When scripts are acquired by knowledge engineers, this can involve col-laboration by system engineers (cf. section 2.5.2). This is particularly important for scripts that contain a significant amount of unexplainable code, such as those involving time management in simulation.
- Whereas more straightforward scripts can be learned by agents, those that reflect the mental models of domain experts are most efficiently modeled by knowledge engineers collaborating with those experts (see section 9.2). However, agents can automatically update such models, as by learning about new therapies for an already-modeled disease from the literature.[22]
- Agents can explain scripts based on:
  - their understanding of the basic shapes of scripts:
    Numbered events occur in order, conditions (introduced by TRY) are ordered if-then statements, and so on. We did not present all script-related conventions here since that would be excessive detail for non-developers;
  - the natural language definitions of the concepts that comprise a script:
    Agents can use definitions directly in explanations or they can semantically analyze them as part of a more reasoning-heavy expla-nation process. Definitions might be missing for scripts that agents learn without the involvement of language;

- if applicable, the natural-language examples in the EXAMPLE field of the concepts that comprise a script; and
- the "EXPL" and/or "EX" fields of procedure calls in scripts.

Later chapters will provide more examples of scripts.

3.3 The Lexicon

The lexicon contains linked syntactic and ontological-semantic descriptions of words and constructions, with the latter covering any combination of words and/or linguistically constrained variables. To start with a simple example, the construction *someone feeds someone*—in the meaning "someone gives food to someone"—is recorded in the LEIA's lexicon as the first verbal sense of *feed*, called feed-v1.

feed-v1
    definition   Someone gives food to someone
    example   Jane fed Fido.
    comments   A different sense covers the ditransitive construction:
                                   Jane fed Fido a steak.
    syntax-type   v-trans
    output-syntax   CL
    syn-struc
      subject $var1
      v $var0
      directobject $var2
    sem-struc
      FEED
        AGENT   ^$var1
        BENEFICIARY   ^$var2

The *definition, example,* and *comments* zones of lexical senses contain human-oriented annotations. The *syntax-type* zone indicates the syntactic construction used in the entry: here, a transitive verb. The *output-syntax* zone indicates the syntactic function of the construction overall: here, a clause.

The *syntactic structure* (*syn-struc*) zone lists the minimal syntactic requirements of the given construction, including the dependency structure, morphological constraints on constituents, and required lexemes, such as the required words in idiomatic expressions. Since feed-v1 is a transitive verb sense, its minimal requirements are the subject, the verb, and the direct object. They are listed in the correct order for the most basic syntactic realization—the active voice.

The *semantic structure* (*sem-struc*) zone expresses the meaning of the construction in terms of ontological concepts, which are written in small caps. This sem-struc is headed by the concept FEED. The carets (^) indicate "the meaning of." So, the AGENT slot is filled by the meaning of the subject, and the BENEFICIARY slot is filled by the meaning of the direct object. The meaning of these arguments can only be computed when an actual input sentence offers words or phrases to fill the variables $var1 and $var2.

Lexical senses can also include *synonyms* and *hyponyms* zones. It is functionally equivalent to record a synonym in its own entry or in the synonyms zone of a different entry. As for hyponyms, it is more informative to record a word in the hyponyms zone of another word than as its own entry because this asserts that the word refers to a subtype of the listed concept. For example, if *pug* is recorded as a hyponym in dog-n1, this makes it clear that *pug* is a kind of dog. If, by contrast, *pug* were to be recorded in its own sense—pug-n1 mapped to DOG—then there would be no way for the agent to know that *pug* is not a generic term for DOG.

The final zone that is needed for some lexical senses is called *meaning-procedures*. It contains calls to procedural semantic routines that either supplement sem-struc descriptions or are fully responsible for the semantic interpretation. An example for which a meaning procedure supplements a sem-struc description is the pronoun *she*. The sem-struc says that *she* means "HUMAN (GENDER female)," and the meaning procedure is needed to identify which female human in the context is being referred to. An example for which a meaning procedure is wholly responsible for the semantic analysis is the adverb *respectively*, as used in *Bears and horses like honey and carrots, respectively*. This meaning procedure must account for the fact that what actually needs to be semantically analyzed is *Bears like honey and horses like carrots*.

Below is the text meaning representation (TMR) for the sentence **Grandfather fed the dog**, which is analyzed using feed-v1.

FEED-1
AGENT	GRANDPARENT-1
BENEFICIARY	DOG-1
TIME	<*find-anchor-time*
lex-sense	feed-v1

GRANDPARENT-1
    HAS-GENDER   male[23]
    lex-sense   grandfather-n1
DOG-1
    COREF   seek-sponsor
    lex-sense   dog-n1
    uses-sense   the-det1

The first frame of the meaning representation naturally looks very similar to the sem-struc of feed-v1, which served as the scaffolding for this analysis. Table 3.3 illustrates this parallelism.

Lexical senses for the other words similarly account for their meanings: *grandfather* and *dog* are plain nouns that are described as "GRANDPARENT (HAS-GENDER male)" in the sense grandfather-n1 and "DOG" in the sense dog-n1. The sense for the determiner *the* indicates that, syntactically, it collocates with a noun and, semantically, it does not have any static meaning; instead, it triggers the procedural semantic routine *seek-sponsor*, which will later attempt to track down its function in the context.[24] The inclusion of a function call indicates that this is not a final TMR; it is the result of an intermediate stage of processing.

Although lexical senses can describe constructions of any form and complexity, the majority reflect standard shapes on both the syntactic and semantic sides. Beginning with syntax, over ninety standard syntactic templates are currently used, whose names fill the *syntax-type* zone of lexical senses. In feed-v1, the syntax-type is v-trans. Table 3.4 shows some additional examples, and the full current list is available in the online appendix.

There are two benefits to asserting syntax types in lexical senses. First, system testing can be selective. If random examples of a particular syntax type are processed correctly, then all examples of that type are expected to work properly—apart from idiosyncratic errors thrown by the data-driven parser.

Table 3.3

The sem-struc descriptions in lexical senses provide the scaffolding for TMRs.

Lexicon		TMR	
FEED		FEED-1	
AGENT	^$var1	AGENT	GRANDPARENT-1
BENEFICIARY	^$var2	BENEFICIARY	DOG-1

Table 3.4
Examples of standard syntax types used in a LEIA's lexicon.

Sense	Syntax Type	Constituents	Output Syntax	Example
refrigerator-n1	n-bare	Noun	N	refrigerator
sleep-v1	v-intrans	Subj sleep$_V$	CL	Lulu slept.
give-v2	v-ditrans	Subj give$_V$ IndirectObj DirectObj	CL	Lulu gave her dog a treat.
nice-adj2	adj-plain	nice$_{ADJ}$ Noun	N	nice weather

Second, the named syntax types can be used as parameters in rules for processing syntactic transformations such as passivization (see section 4.2.2).

All lexical senses are also labeled with their *output-syntax*, which is the type of constituent they create. For example, since adjectives are described in conjunction with the noun they modify, their output-syntax is N (noun); and since verbs are described with their arguments, their output-syntax is CL (clause). The value of output-syntax asserts how the given lexical sense can participate in larger constructions.

Many standard constructions include optional elements such as optional arguments (e.g., the direct object of *read*) and adjuncts. Adjuncts are listed in lexical senses when they are particularly common and when listing them will help in disambiguation. For example, many adjuncts are headed by prepositions, which are multiply ambiguous. Asserting what they mean in a particular construction is not only helpful; we think it mimics people's knowledge of constructions. For example, the "fasten" sense of the verb *secure* is often used with a prepositional phrase headed by *with* to express the INSTRUMENT: *Fred secured the tent with stakes.*[25] Adding such information to lexical senses boosts the agent's power of disambiguation. This is a good example of a low-cost, high-payoff strategy in the overall process of knowledge acquisition (cf. chapter 9).

There are also nonstandard constructions, whose value for output-syntax is *atypical*. They can include any number and type of ordered constituents. For example, the semantically vacuous expression *"The thing is, is that* Clause" is recorded in the lexicon as this specific sequence of words followed by a clause of any shape. Nonstandard syntactic constructions can have many different kinds of output-syntax. They need to be tested separately to determine how they will be treated by the parser.

The tidy inventory of syntactic construction types in the lexicon belies the massive complexity of language that must be handled during language understanding. For example, any construction that includes a noun phrase needs to accommodate any shape of noun phrase, such as:

- a car
- a nice, expensive car
- my friend's nice, expensive car
- my friend's nice, expensive car that she got from her parents as a graduation present
- my friend's nice, expensive car that she got from her parents as a graduate present and has been driving to the beach every day all summer.

The lexicon includes only most basic uses of words, such as the active forms of verbs and the attributive uses of adjectives. This is sufficient because the language analyzer can handle generativity using a model of transformation processing that is psychologically plausible and computationally practical—as explained in section 4.2.2. This generative approach offers practical benefits in terms of knowledge acquisition and maintenance:

- Most words have multiple senses, defined as particular correlations of syntactic and semantic elements. Light verbs, such as *have*, *take*, and *make*, have dozens of senses each. If all of these senses were listed in all of their possible shapes—including passive, imperative, and participating in every type of question—the size of the lexicon would increase dramatically.

- Non-basic uses of constructions can combine, leading to further combinatorial explosion. For example, *Fido, he was fed by the girl who was recently hired as his dog sitter* involves subject dislocation, passivization, and a relative clause construction.

- If all of these non-basic uses and combinations thereof were listed explicitly, then every time a lexical sense was edited, all of the associated senses would need to be edited. This will not be a rare occurrence as the ontology grows over time. For example, if a knowledge engineer decides to split a more coarse-grained concept, like RUN, into multiple children, like JOG and SPRINT, then all associated lexical senses will need to be remapped: jog-v1 will remap from RUN to JOG, sprint-v1 will remap from RUN to SPRINT, and so on.

So far, we have discussed the syntactic side of lexical descriptions. Turning to semantics, descriptions of word senses also have more and less typical forms. Table 3.5 shows some typical forms of sem-strucs.

Although there are many typical shapes of sem-strucs, it is not practical to try to list them, nor is it needed. (Recall that the main reasons for formally classifying syn-strucs were (a) to align them with the possibly unpredictable output of the external parser and (b) to anticipate how constructions can interact with each other.) Instead, it is better to conceptualize semantic descriptions as a generative process. At the highest level, the legal form and content of sem-struc descriptions is as follows:

- They can include any number of frames.
- Each frame can be headed by a concept, a variable, or a set indicator.[26]
- Each frame can include any properties that are appropriate for its head type.
- Property fillers can, themselves, be frames; that is, constituents can be nested.
- The sem-struc zone of a lexical sense can be empty. In some cases, this is because a word has no meaning—as for the disfluency markers *uh* and

Table 3.5
Examples of typical sem-struc zones of lexical senses in a LEIA's lexicon.

Description	Sense	Example of Sem-Struc
Concept	refrigerator-n1	REFRIGERATOR
Concept with one property	sleep-v1	SLEEP EXPERIENCER ^$var1
Concept with two properties	throw-v1	THROW AGENT ^$var1 THEME ^$var2
Variable with one property	blue-adj1	^$var1 COLOR blue
Multiple frames of the any type	must-v1	MODALITY TYPE OBLIGATIVE VALUE 1 SCOPE ^$var2 ^$var2 AGENT ^$var1

er. In other cases, an empty sem-struc reflects the fact that the entity has no static meaning. As mentioned earlier, the adverb *respectively* in inputs like *Bears and horses like honey and carrots, respectively* triggers a procedure that recasts the sentence as the meaning of *Bears like honey and horses like carrots.* So, the entire semantic interpretation of the word *respectively* is procedural.

The LEIA's English lexicon currently contains around fifteen thousand senses using the simplest counting method: the number of listed senses (e.g., feed-v1, perform-v6). However, the actual coverage of the lexicon is much greater because simple counting does not account for:

- the large number of synonyms and hyponyms recorded in lexical senses;
- the selection or non-selection of optional elements;
- transformations, and combinations thereof, that make the lexicon generative;
- the productive handling of numbers and named entities; and
- lexicon-wide processes of derivational morphology.

Our point in citing a number at all is to show that we are trying to fundamentally solve the problems of natural language understanding and generation, and this requires handling lexical ambiguity and paraphrase. So, although the LEIA's lexicon currently contains nowhere near human-level lexical knowledge, it includes extensive polysemy and synonymy, which creates a rigorous testbed for the agent's natural language understanding and generation systems.[27]

Although most of our recent work has involved English, both the approach and much of the knowledge substrate—even the lion's share of the lexicon—are language-independent. The reason why LEIA-style lexicons can be ported across languages is because the most difficult part of lexical acquisition is describing semantics, both static (recorded in the sem-struc zone) and procedural (recorded as function calls in the meaning-procedures zone). So, creating a lexicon of French or Russian from the existing English one primarily involves changing the words used to convey the given meaning. If any syntactic or semantic tweaks are needed, they are typically quick and simple.[28] It is noteworthy, in this regard, that the theory of Ontological Semantics that underpins LEIA language processing has its roots in interlingual machine translation.[29] Section 5.4 gives a taste of the crosslinguistic applicability of our approach to language understanding using evidence from Russian.

3.4   The Opticon and Analogous [Sense]icons

Just as the lexicon supports the translation of language inputs into onto-logical concepts, so, too, must analogous knowledge bases for other chan-nels of perception. To date, we have worked only with an opticon, but it is straightforward to apply the approach to a hapticon (for touch), olfacticon (for smells), physiocon (for sensor-detectable features of human physiol-ogy), and so on.[30] Focusing on the opticon, the entry for any object, event, or scene includes:

- a head that is a set of one or more visual representations (static images or video clips) that serve as exemplars;
- a visual representation of the components of the object, event, or scene, along with their spatial relations and links to the components' own opti-con entries;
- a meaning procedure that helps the agent to recognize the object, event, or scene and its parts; and
- a meaning procedure that helps the agent to recognize the individual optical features that distinguish the object, event or scene.[31]

Whereas it is self-evident why an embodied agent would need to be able to detect things like a stop sign (using vision), a red-hot surface (using touch), or something burning (using smell), the utility of physiological fea-ture detection deserves further comment.

Human performance on a task can be affected by the person's physi-cal, emotional, and cognitive states. When humans collaborate with each other, they naturally respond to behavioral manifestations of such states: for example, teachers give hints to students who are frustrated, workers lend a hand to teammates who are exhausted, and supervisors offer reas-surance to subordinates who are overwhelmed. In order for agents to serve as reliable collaborators, they, too, must be able to detect and appropriately respond to people's physical, emotional, and cognitive states.

Human behavior research has discovered correlations between measurable physiological features—such as heart rate variability, electrodermal activity, pupil size, and eye movements—and states such as arousal, engagement, stress, fear, mental effort, and physical exertion.[32] Making use of such features in cognitive systems involves:

- developing a physicon that maps sensor outputs to ontologically grounded feature values;
- developing a dedicated recognition module that perceives physiological inputs and interprets them according to the physiocon;
- developing a dedicated interpretation module that contextually interprets physiological features in terms of people's physical, emotional, and cognitive states: for example, in a given context, increased heart rate might be explained by stress, physical exertion, or exposure to heat; and
- developing reasoning functions that guide the agent in responding to different human states depending on a large number of features of the context including the respective roles that the human and agent are playing, the type of application, the application domain, and so on.

Returning to the overall topic of this section, just as LEIAs need a lexicon to map between language and ontologically grounded meanings, they need analogous knowledge bases to map between other channels of input and ontologically grounded meanings. It is these meanings that agents use as input for reasoning about action.

3.5   Episodic Memory

Episodic memory includes stored information about instances (exemplars) of ontological concepts as well as meaning representations that the agent generates during its operation. Episodic knowledge structures include timestamps, provenance, and other relevant metadata.

The agent's episodic memory is divided into spaces, which is a common practice in memory management. Memory spaces allow the agent to rapidly access sets of known instances that share a common category or purpose. For example, one part of an agent's episodic memory contains a fact repository of the agent's beliefs about entities it knows—for example, the capital of Belgium and the hair color of its human collaborator Ben. Another part of the agent's episodic memory contains information about its past successes and failures at carrying out a particular kind of plan (recall that plans are instances of ontological scripts). Memories of plans allow the agent to, in certain cases, bypass detailed decision-making about action and, instead, carry out a reflexive action. Specifically, when the agent needs to decide which plan to use to achieve a goal, it can search its episodic memory for past cases when particular plans successfully achieved the goal. It can then compare its current situation model to the situation models associated with the successful plans and select the best match. The agent can then instantiate another copy of that plan in anticipation that it will work as well as the last time. This kind of operation is an example of case-based reasoning.

The content of the agent's episodic memory is made available to all reasoning heuristics throughout the system, allowing any algorithm to inspect what the agent knows, what it has recently encountered, what it is currently thinking about, and what is on its agenda. As with ontological knowledge, episodic memory is indexed in a variety of ways—by relevant domain space, but also by type, timestamp, and more. Operations devoted to consolidation and other updates of the episodic memory, as well as the way the agents model forgetting, are outside the scope of this book.

4   Language Understanding and Generation

True to their name, Language-Endowed Intelligent Agents count language as a top priority. Like the humans they emulate, LEIAs learn, collaborate, and explain using language, so their language-oriented capabilities could not be more important. Many of those capabilities are described in our recent book, *Linguistics for the Age of AI* (2021), and will not be recounted here. Instead, this chapter does this following:

- It provides the minimum necessary background about language understanding to make this book self-sufficient. All examples of dialog, learning, and explaining in the upcoming chapters rely on natural language understanding.
- It reports select advances in the LEIAs' language understanding module since the publication of *Linguistics for the Age of AI* by way of illustrating how LEIA microtheories evolve over time.
- It presents the new natural language generation module used by LEIAs.
- It juxtaposes the microtheory of *construction semantics* used by LEIAs with the human-oriented theory of *construction grammar* and shows how LEIA modeling fulfills the more rigorous demands of computation.

The inevitable reliance of this chapter on material from *Linguistics for the Age of AI* underscores that the LEIA program of R&D can neither be encapsulated in a single monograph nor frozen at a moment in time. So, although we attempted to make this book largely free-standing, *Linguistics for the Age of AI* is highly recommended as a companion read.

4.1   Introduction

To serve as intelligent collaborators, LEIAs need to be able to understand and generate meaningful natural language.

Language understanding involves translating natural language inputs—which can be ambiguous, elliptical, ill-formed, and implicature-laden—into fully specified, ontologically grounded text meaning representations (TMRs) that convey their intended meaning. The global interpretation of an input is built up from the interpretations of progressively larger groups of words, phrases, and constructions, informed by features of the context.

Language generation, for its part, involves converting ontologically grounded generation meaning representations (GMRs), which result from an agent's reasoning about action, into contextually appropriate natural language utterances.

As described earlier (section 2.1.2), TMRs and GMRs are very similar apart from their metadata. Moreover, many of same linguistic and extralinguistic challenges confront both language understanding and generation, albeit in different guises. For example:

- The language understander must disambiguate polysemous words, whereas the language generator must select among available paraphrases.
- The language understander must resolve referring expressions whereas the language generator must choose among options for referring to entities.
- The language understander must reconstruct elided material, whereas the language generator must decide when to produce elliptical utterances.
- The language understander must interpret indirect speech acts, whereas the language generator must decide when to use them.

Language understanding and generation mirror each other on the input and output flanks of the agent architecture, as shown by figure 4.1, which is a language-specific version of the generic architecture in figure 2.1.

The language-oriented path through the architecture differs from the generic architecture as follows:

- The input is speech or text.
- The generic module Perception Recognition is realized as Language Recognition.
- The generic module Perception Interpretation is realized as Language Understanding, which yields a text meaning representation (TMR).
- Action Specification yields a GMR (generation meaning representation), which is specific to language generation.

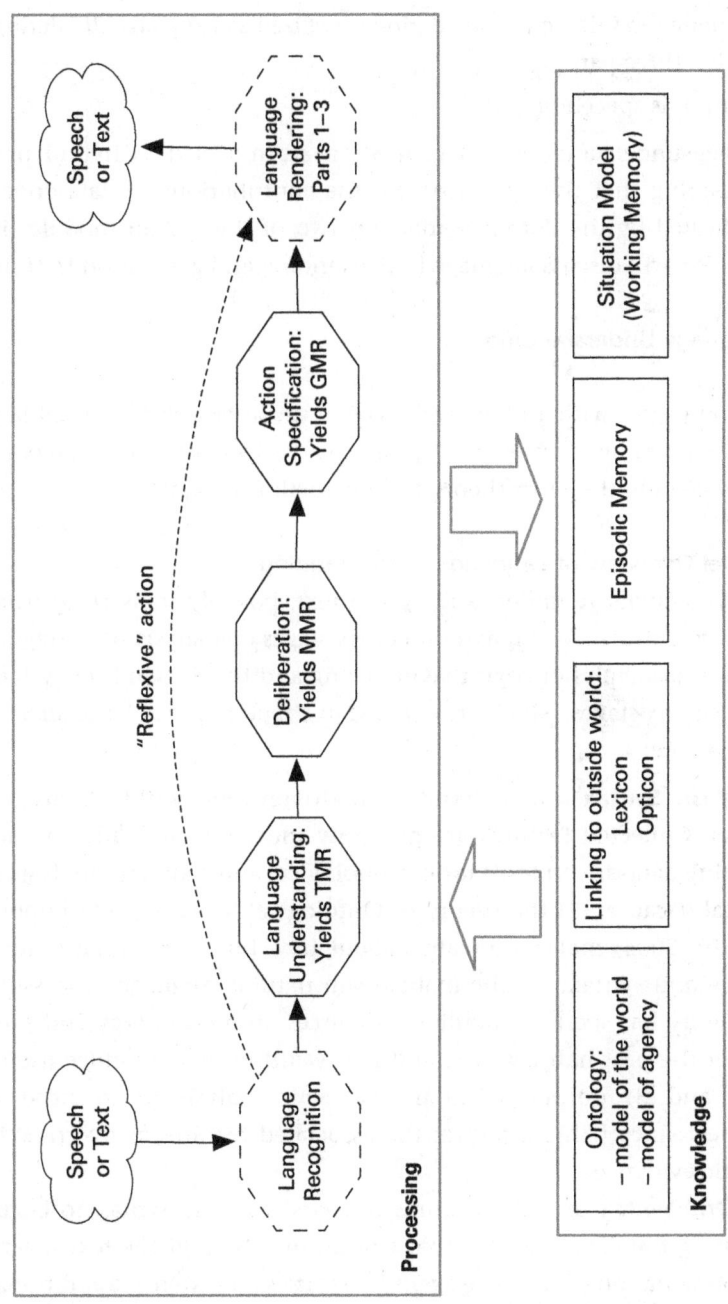

Figure 4.1
Natural language understanding and generation reflect a specialization of the generic architecture shown in figure 2.1.

- The generic module Action Rendering is realized as Language Rendering, which is a three-part process.
- The output is speech or text.

Language understanding and generation are modeled as hybrid processes, meaning that they benefit from the contributions of data-driven tools, indicated by the dotted borders on two of the system modules in figure 4.1. We will discuss language understanding and generation in turn.

4.2 Language Understanding

This section begins with a high-level recap of the approach to natural language understanding reported in *Linguistics for the Age of AI* and then reports recent advances in the microtheory of construction semantics.

4.2.1 Brief Overview of Language Understanding

The input to language understanding is a text, possibly transcribed from speech, which undergoes a maximum of six stages of analysis. This staging reflects both principles of cognitive modeling and the demands of system building. The six stages, which are detailed in chapters 3–7 of *Linguistics for the Age of AI*, are:

Stage 1. **Basic Syntax** runs external, data-driven tools within a custom wrapper. Collected features are primarily morphological and syntactic.[1] Using data-driven tools for morphological and syntactic analysis is practical because: (1) the theory of Ontological Semantics (Nirenburg & Raskin, 2004) makes no claims about how humans compute morphology and syntax; (2) the mainstream natural language processing community has spent tremendous resources on morphology and syntax over the past half century, and the resulting tools produce useful results; and (3) morphological and syntactic analysis do not need to be perfect since LEIAs interpret the associated features as overridable heuristic evidence.

Stage 2. **OntoSyntax** uses the features collected in Basic Syntax to begin populating the data structure that will ultimately hold the text meaning representation (TMR). The agent also carries out extensive additional analysis of the input, which involves such things as selecting which lexical senses might be used to analyze the input, treating syntactic transformations, and initiating the learning of new word senses.

Stage 3. **Basic Semantics** focuses on lexical disambiguation and the establishment of the semantic dependency structure for individual sentences. Typically, this stage produces multiple candidate TMRs, which are improved upon and, ideally, pruned down to a single one by the end of stage 6.

Stage 4. **Lexically Triggered Procedural Semantics** runs specialized routines that are called from the meaning-procedures zone of lexical senses. These cover many different kinds of phenomena, from coreference (*she*) to scalar modifiers (*very hot*) to words whose meanings guide the interpretation of a whole utterance (*respectively*).

Stage 5. **Extended Semantics** attempts to resolve residual ambiguities, incongruities, and underspecifications using:

a. additional knowledge bases, such as ontologically grounded repositories for analyzing nominal compounds and metonymies;

b. additional algorithms, such as the ones that attempt to specify ungrounded comparisons (*X is better*—than what?) and understand indirect modification (*a fast road* is a road where one *can drive* fast); and

c. a multi-sentence window of discourse, along with analysis methods that exploit it.

Stage 6. **Situational Reasoning** invokes all of an agent's abilities—such as plan-and-goal-oriented reasoning, image analysis, and mindreading of the interlocutor—in an attempt to arrive at a single, precise, contextually appropriate, high-confidence analysis of a language input.

Each stage of language analysis results in a knowledge structure that the agent can use to reason about *actionability*. It asks itself, *Is the nascent TMR, in its current state, adequate input to reasoning about action?* For example, a LEIA that is tasked with fundamentally analyzing only inputs about a particular topic can, after only one or two stages of analysis, determine whether the input contains words or concepts of interest; if not, its action can be to ignore it. Such skimming is useful for learning by reading and for agents exposed to off-topic chitchat by their human collaborators. Similarly, an agent tasked with learning new words by reading might focus on the results of Basic Semantics, opting to exclude any sentences that require coreference resolution or other advanced reasoning.[2]

One way to think about language understanding is in terms of knowledge used and knowledge produced. LEIAs use both prerecorded (static) knowledge and knowledge generated during system operation. This reflects the following principle of content-centric cognitive modeling: decision heuristics within any perception or reasoning module may require knowledge from any of the system's static knowledge resources and/or any of the dynamic knowledge generated by any of the system's processing modules. Table 4.1 shows how this principle plays out in language understanding.

All of this knowledge is inspectable, as are traces of which specific elements of knowledge the agent uses in each component function.

A strategic decision is to not permit backtracking in language understanding: the agent populates the analysis space with all possible candidates and prunes out candidates throughout the stages of analysis. We have not yet modeled how an agent might realize, later on, that its interpretation of a previous input was incorrect. Whatever the details of that model might be, it will still likely not involve backtracking. Instead, the agent will reanalyze the input using the newly available information.

On choices and priorities: Incremental processing or not?

The language understanding system described in *Linguistics for the Age of AI* processed input incrementally in an attempt to mimic people's interpretation of language as it unfolds. However, the implementation of incrementality was guided more by engineering considerations than by cognitive modeling, and LEIAs ended up computing more candidate subsentential analyses than a person would. Our recent work has de-emphasized incrementality for two reasons. First, it remains a research issue when and to what degree incremental processing might enhance agent operation since natural speech is quite fast and the ends of utterances are reached quickly. Second, at the current stage of agent development, the large research effort needed to build an expectation-driven model of incrementality is not a top priority. Accordingly, we will not address incrementality further in this book.

4.2.2   Recent Advances in Construction Semantics

Construction semantics comprises the first four stages of language understanding, during which the syntactic structure of an input informs its semantic interpretation. This book includes a deep dive into construction semantics for two reasons. First, LEIA development is cognitive science applied to AI and, as such, is informed by and informs various theories of cognition. Reporting such influences—in this case, related to the human theory of construction grammar—is an important part of the scientific enterprise. Second, a big challenge in the computational modeling of language understanding is developing an integrated sequence of methods for

Table 4.1
Sample knowledge used and produced by the stages of language understanding.

Stage	Knowledge Used	Knowledge Produced
Basic Syntax	Whatever the external, data-driven engines use, such as: – a lexicon – a gazetteer – a syntactic and/or morphological grammar – an annotated corpus	– sentence boundaries – word information: lemmas, part of speech tags, morphological features – constituency parse – dependency parse – named entity information
OntoSyntax	– all of the knowledge available & produced thus far – syntactic descriptions in the lexicon – syntactic transformations – disfluency recovery rules – parse reambiguation rules – rules for linking lexical senses to words of input – rules for combining lexical senses in syntactically valid ways – rules for scoring sense combinations – rules for positing new word senses	– candidate linkings between elements of input and senses in the LEIA's lexicon – candidate combinations of lexical senses – nascent lexical senses for unknown words (full syntax, underspecified semantics)
Basic Semantics	– all of the knowledge available & produced thus far – the semantics of the word senses in the candidate solutions – syntax-to-semantics linking rules derived from lexical senses – ontological constraints to gauge the confidence of semantic dependencies – microtheories of linguistic phenomena, including modality, aspect, mood, and time.	– word-sense disambiguation decisions – semantic dependency decisions – identification of the availability of direct and indirect speech-act interpretations – a more precise specification of the semantics of unknown words

(continued)

Table 4.1
(continued)

Lexically Triggered Procedural Semantics	– all of the knowledge available & produced thus far – microtheories implemented as procedural semantic routines for the treatment of specific linguistic phenomena, such as coreference and scalar modifiers (e.g., *very* tall).	– scored candidate sponsors for overt and elided referring expressions – explanations of non-coreferential definite descriptions, including bridging references – the results of other kinds of lexically triggered procedural semantic analysis
Extended Semantics	– all of the knowledge available & produced thus far – microtheories of nonliteral language processing: e.g., metaphor, metonymy – a repository of concept-based nominal compounding constructions – a repository of classes of metonymies – microtheory of extra-clausal disambiguation – microtheory of integrating fragments into the discourse context	– the semantic/pragmatic incorporation of fragments into the discourse context – a culled set of candidate text meaning representations thanks to additional methods for resolving ambiguity and incongruity
Situational Reasoning	– all of the knowledge available & produced thus far – goal and plan inventory – agenda of active goals and plans – models of self and other agents – microtheory of mindreading – microtheory of indirect speech acts – results of nonlinguistic perception interpretation – active concept instances in the situation model – the microtheory of reference (i.e., anchoring referents to agent memory) – rules for scoring candidate analyses	– the intended, context-sensitive meaning of the input – true reference resolution (to memory) – improved interpretation of unknown words – recognition of any residual ambiguities or underspecifications – confidence score for each candidate analysis

treating the full scope of linguistic phenomena. Here, we provide explicit guidance to developers who might choose to use construction semantics in developing their own LEIA systems.

OntoSyntax The first cognitively modeled stage of language understanding is OntoSyntax, Basic Syntax being carried out by imported, data-driven tools. OntoSyntax not only adapts the syntactic parse to make it compatible with a LEIA's knowledge bases, it also carries out substantial additional syntactic processing. That processing is divided into four steps, which we describe in turn: (1) building the working-memory lexicon, (2) reambiguating premature parsing decisions, (3) processing transformations, and (4) carrying out SynMapping.

Step 1: Building the working-memory lexicon When the agent begins to process a sentence, it creates a working-memory lexicon (WMLexicon) that contains copies of the word senses from the base lexicon that might be needed for the analysis. These copies can be modified as part of sentence processing without perturbing the base lexicon. The WMLexicon is created as follows:

1. For each word of input, all senses in the lexicon that have the needed part of speech are copied into the WMLexicon.

2. If a word of input is syntactically unexpected, a new sense is generated for it. Words are syntactically unexpected if they are (a) completely absent from the lexicon, (b) attested but in the wrong part of speech, or (c) attested in the correct part of speech but with the wrong syntactic dependencies—for example, the input might use the transitive sense of *walk*, as in *She walked the dog*, whereas the lexicon includes only an intransitive sense. This sense-generation function also creates senses for named entities, mapping them to their appropriate ontological concepts. For example, human names are mapped to HUMAN with the property HAS-NAME being filled by the person's name.

3. If the input contains any words that might be semantically null due to their participation in an idiomatic expression, a null-semantics sense is generated for them. For example, when *bucket* is used in the idiomatic expression *kick the bucket*, it does not carry any independent meaning, so none of the senses of *bucket* recorded in the lexicon will fit. A null-semantics sense plans for this eventuality and allows the word to be treated like any other in the generic algorithms for syntactic and semantic analysis.

The lexical senses that are dynamically generated by the second and third processes above are added to the WMLexicon, completing the solution space for analyzing the sentence.

Step 2: Reambiguating premature syntactic decisions   Syntactic parsers are forced to make certain kinds of decisions that are actually semantic in nature, which means that they are set up to fail. This function detects such situations and reintroduces the ambiguity so that the semantic analyzer can weigh in later on. Three cases are currently covered:

- **PP (prepositional phrase) attachments.** When a PP immediately follows a post-verbal NP, it can modify either the verb or the adjacent NP. A famous example is *I saw the man with the telescope.*
  - If the telescope is the instrument of seeing (it is being used by the speaker to see better), then the PP attaches to the verb: I [$_{VP}$ saw [$_{NP}$ the man] [$_{PP}$ with the telescope]].
  - If the telescope is associated with the man (he is holding or using it), then the PP attaches to the NP: I [$_{VP}$ saw [$_{NP}$ the man [$_{PP}$ with the telescope]]].
- **Nominal compounds.** Nominal compounds containing more than two nouns have an internal structure that must be analyzed semantically. Compare:
  - [[kitchen floor] cleanser]
  - [kitchen [floor lamp]]
- **Phrasal verbs.** In English, many prepositions are homographous with verbal particles. Consider the collocation *"go after* NP," which can have two different syntactic analyses associated with two different meanings:
  - When *go after* is a phrasal verb, the expression has the idiomatic meaning *pursue, chase*: *The cops went after*$_{Particle}$ *the criminal*$_{DirectObject}$.
  - When *go* is used compositionally with the prepositional phrase *after* NP, it has the meaning *do some activity after somebody else finishes their activity*: *The bassoonist went*$_{V}$ *after*$_{Prep}$ *the cellist*$_{ObjectOfPrep}$.

While there are clearly two syntactic analyses of *go after* that are associated with different meanings, and while there is often a default reading depending on the meanings of the subject and object, it is impossible to confidently select one or the other interpretation outside of context. After all, *The cops went after the criminal* could mean that the cops provided testimony after the criminal finished doing so, and *The bassoonist went after the cellist* could mean that the former attacked the latter for having stepped on her last reed.

In all of these situations, LEIAs reambiguate the parse, making additional candidate analyses available.

Step 3: Processing transformations   The lexicon contains only the most basic syntactic realizations of argument-taking words, such as the active voice of verbs (cf. section 3.3). The transformation processor prepares for the large number of non-basic ways that argument-taking words can be used in language. For example, the lexical sense for the transitive use of the verb *feed* expects a subject and a direct object, in that order: *Jane fed Fido*. However, *feed* in this meaning can be used in many other syntactic configurations, such as those illustrated in table 4.2.

The fact that verbs can be used in many ways in sentences is best accounted for by procedures, called *transformations*, that dynamically align non-basic word uses with the basic shapes recorded in the lexicon.

For LEIAs, transformations are not only syntactic; they are semantic as well. Whereas the term *transformation* originates in syntactic theory, for LEIAs, transformations involve syntactic structures and their linked semantic interpretations. Specifically:

- Transformations modify the syntactic expectations of the basic word sense recorded in the syn-struc of a lexical sense.
- The semantic constraints on the constituents, recorded in the sem-struc, are retained.
- Dependencies among the constituents are retained or modified, as applicable.
- If applicable, calls to procedural-semantic routines, which are recorded in the meaning-procedures zone of the original lexical sense, are retained or modified.
- Semantic and/or discourse features contributed by the transformation are added. For example, the passive voice adds the feature "DISCOURSE-STATUS TOPIC" to the original direct object that was promoted to the subject position.

Note that the word *transformation* can refer to both the *process* of transforming (the algorithm implemented as code) and the *result* of that process. Both meanings are relevant for LEIAs.

Transformations fire when particular syntactic triggers are identified in the parse, and they modify the appropriate word sense in the WMLexicon. For example, if the input uses a verb in the passive voice, the lexical sense for the verb—which is in the active voice—is transformed into the passive voice, leaving no trace of the original, as shown in table 4.3.

Table 4.2
Sample non-basic uses of *feed* in the sense *give someone food*.

Transformation type	Example
Passive voice	Fido was fed (by Jane).
Subject dislocation	Jane, she fed Fido.
Topicalization	It was Jane who fed Fido.
Xcomp	Jane wanted to feed Fido.
Verb phrase coordination	Jane walked Fido and then fed him.
Yes-no question	Did Jane feed Fido?
Question using wh-adverb	When did Jane feed Fido?
Question using wh-adverb and passivization	When was Fido fed?

Table 4.3
Example of a transformation: Passivization of transitive *feed*.

feed-v1 basic		feed-v1-passive-trans	
definition	to give food to	definition	–
example	Alice fed Fido	example	Fido was fed by Alice
syntax-type	v-trans	syntax-type	passive-trans
output-syntax	CL	output-syntax	CL
syn-struc		syn-struc	
subject	$var1	subject	$var2
v	$var0	aux	$var3 (root be)
directobject	$var2	v	$var0 (form past-part.)
sem-struc		pp (opt+)	
FEED		prep	$var4 (root by)
AGENT	^$var1	obj	$var1
BENEFICIARY	^$var2	sem-struc	
		FEED	
		AGENT	^$var1
		BENEFICIARY	^$var2
		^$var2	
		DISCOURSE-STATUS TOPIC	
		^$var4 null-sem+	

Passive Transformation

Triggering features: a psubj dependency and/or the constituents "NP *be* [past participle of a verb that takes a direct object]"

Added feature(s):   The passive subject is assigned the feature "DISCOURSE-STATUS TOPIC"

Example: Alice fed Fido. → Fido was fed (by Alice).

This transformation can be triggered by a psubj (passive subject) dependency in the syntactic parse and/or the sequence of constituents "NP *be* [past participle of a verb that takes a direct object]." It is important to formulate the latter condition as "a verb that takes a direct object" rather than "a transitive verb" because this transformation applies to more than just basic transitive verbs—it also applies to optionally transitive verbs, ditransitive verbs, transitive verbs with any number of obligatory or optional prepositional phrases, verbs that take a direct object and an xcomp (*He encouraged her to go → She was encouraged to go by him*), and so on. This transformation remaps case roles to the appropriate constituents of the input and adds the feature "DISCOURSE-STATUS TOPIC" to the meaning of the passive subject.

Table 4.4 shows some more examples of transformations. For each one, the agent is supplied with a function that maps the basic lexical sense into the transformed one in a way analogous to the passive example above.

This is just a subset of the full inventory of transformations used in English, but it is sufficient to establish a basic approach to transformations that can be applied to all others.

An important feature of our approach to transformations is their combinability. For example, *When was Fido fed?* involves both passivization and a wh-question transformation. The question is, how best to handle such combinations? Should they be treated as a complex whole—that is, processed using a single rule—or should individual transformations be dynamically combined on the fly?

Our solution is inspired by theories of human cognition that consider frequency effects and economy of effort. We hypothesize that people store frequent combinations of transformations as construction-like entities that can be looked up rather than recomputed. This minimizes the cognitive load of processing such inputs. Modeling LEIAs to do the same involves recording combined transformations covering frequent combinations like "wh-adverb & passivization." This facilitates processing inputs like *When was Fido fed?*

But that still leaves the matter of infrequent combinations, which need to be dynamically computed. For LEIAs, this means identifying which transformations are needed and applying their effects in a particular order. Our current approach is to order the application of transformations according to the constituent size that they operate on: NP-level transformations apply first, then basic argument-realization transformations, then basic clause-level transformations, then question-oriented transformations, and finally

Table 4.4
More examples of transformations.

Basic argument-realization transformations

Indirect object passivization	The man was granted a pardon by the judge.
Imperative	Ask the driver for his keys.
Indirect object realized as "*to* NP"	The boy gave a flower to the girl.
	[The lexicon has "Subj give IndirectObj DirectObj"]

Question-oriented transformations

Yes-no question	Did you see Henry?
Direct object as a wh-question	Who did he see?
Subject as wh-question	Who saw it?
Indirect object as a wh-question	Who did you give the basket to?
Wh-adverb	When did Alice feed Fido?

Conjunction-oriented transformations, which cover any number of conjoined entities[3]

NP-conj expansion	He adopted a cat, a dog, and a bird.
N-conj expansion	She bought many pillows and vases.
VP-conj expansion	They ordered a pizza and played chess.

Topicalization strategies

It-was topicalization	It was Jane who fed Fido.
As-for topicalization	As for Jane, she fed Fido.
As-far-as topicalization	As far as Fido is concerned, Jane fed him.

Miscellaneous transformations

Present participle construction	Painting landscapes, Penelope is happy.
Perfect participle construction	Having painted a landscape, Penelope took a nap.
Subject dislocation	The witch, she cast a spell over him.
Direct object fronting	Pizza, I like.
Gerund subject	Eating cake is nice.
Prep-part[4] ordering	They whisked the children away.
	[The lexicon has "Subj whisks away DirectObj."]
Preposition swapping[5]	They absolved him from his crimes.
	[The lexicon has *of* not *from*.]
Elided verb reconstruction	James swam yesterday but can't today.
	[*Today* needs a reconstructed verb to modify.]
Elided noun reconstruction	I prefer the green.
	[*Green* needs a reconstructed noun to modify.]
Cross-sentential conjunction use	I like it. But he feels differently.
	[During stages 1–4 of language understanding, sentences are processed individually, which would leave sentence-initial conjunctions without their first conjunct if there was no transformation.]
Relative pronoun use	I saw the picture she drew.
	[*The picture* is the understood direct object of *drew* as well as the direct object of *saw*.]

---

extra-clausal transformations. (This is a formal classification that differs from the conceptual one presented above.) As an example, the following steps are needed when using the basic ditransitive sense of *give*, "Subj give IndirectObj DirectObj," to analyze the sentence *Was the doughnut given to Sally?*

1. [Basic argument-realization transformation] The indirect object is realized as "*to* NP":

Subj$_1$	*give*	IndirectObj$_2$	DirectObj$_3$	→ Subj$_1$ *give*	DirectObj$_3$	to	NP$_2$
(s.o.)	gave	Sally	the doughnut	→ (s.o.) gave	the doughnut	to	Sally

2. [Clause-level transformation] Passivization:

Subj$_1$	*give*	DirectObj$_3$	to	NP$_2$	→ Subj$_3$		*be*$_{Aux}$	*given*	to	NP$_2$
(s.o.)	gave	the doughnut	to	Sally	→ the doughnut		was	given	to	Sally

3. [Question-oriented transformation] Yes-no question formation:

Subj$_3$		*be*$_{Aux}$	*given*	to	NP$_2$	→ *Be*$_{Aux}$	DirectObj$_3$		*given*	to	NP$_2$?
the doughnut		was	given	to	Sally	→ Was	the doughnut		given	to	Sally?

Even if this particular combination of transformations turns out to be frequent enough to be stored as a single, complex transformation, a productive process like this is needed for other cases. It is a research question whether any single algorithm can correctly account for all possibilities or whether—as is so common in language—there are idiosyncratic aspects of how transformations combine that must be accounted for by explicit rules.[6]

As mentioned earlier, transformations modify the senses in the WMLexicon in particular ways, preparing them to match a specific input. The use of transformations makes the LEIAs' lexicon generative, allowing the stored lexicon to remain compact and streamlined. However, it is worth mentioning that having a streamlined lexicon that is dynamically expanded by transformations is not the only option. In fact, transformations are explicitly rejected in at least some varieties of construction grammar (cf. section 4.4). An alternative to dynamic transformations is having a vastly larger lexicon that explicitly prepares for all uses of words and multiword expressions. This would mean, for example, that each verb sense would need to be provided with not only the active voice but, as applicable, also the passive voice, the imperative, the verb used in all manner of questions, and so on. Since each verb can have dozens of senses, and each would need to be expanded accordingly, this would result in a massive increase in the size of the lexicon and a considerable burden for the humans in the loop of lexical acquisition and system testing. A third option would be to not record all of these options statically but, instead, dynamically expand the lexicon along transformational lines prior to runtime. Ultimately, the choice between just-in-time transformations and a pre-runtime lexical expansion comes down to engineering considerations, such as whether processing speed is affected by the different strategies.

To conclude this section on transformations, we must emphasize that accounting for the full transformational complexity of natural language on both the syntactic and semantic sides will require further development work, and it is imperative to use corpus evidence as a guide. It would be impractical—and, indeed, theoretically questionable—to spend effort providing for combinations of transformations that people do not actually produce.[7]

Step 4: SynMapping   The goal of syntactic mapping, SynMapping, is to create syntactically valid combinations of lexical senses that are candidate solutions for interpreting an input. SynMapping matches features of the input parse to word senses in the WMLexicon, which might have already been modified by transformations. If a word sense's features are incompatible with the input, that sense is excluded from the candidate space. For example, if a verb sense requires a direct object but the input does not contain one, that sense is rejected.

True to its name, SynMapping focuses on syntax—the syntactic parse of the input along with the syn-struc zones of lexical senses. However, as with syntactic transformations, SynMapping must carry along with it the semantic sides (sem-struc zones) of those lexical senses.

SynMapping covers grammatical inputs, not the kinds of irregular, disfluent, and fragmentary utterances that are common in spontaneous speech. For example, in response to being asked to pass the spatula, someone might say, "I don't see . . . wait, um, this one?" Such outliers are dealt with in two ways:

1. After the initial parse, disfluencies are stripped and the input is reparsed, which often results in a canonical syntactic structure that can be processed in the normal way.

2. If, after disfluency stripping, the input is still non-canonical, the agent abandons construction semantics, which uses syntax to inform semantics, and jumps to stage 6 of language understanding, during which it attempts to analyze semantics directly, assembling candidate meanings in the most reasonable way based on expectations recorded in the ontology.[8]

The challenge of SynMapping is managing the potentially very large number of candidate combinations. Our solution, which should be of interest to developers, is presented in the online appendix. For general readers, two points are important. First, the agent *does* manage to create all and only the syntactically viable combinations of word senses that could be used to analyze the input. Second, we achieved this not using computational strongarming but, instead, using a model developed through collaboration between a knowledge engineer and a software engineer. The reason to insert a knowledge engineer into a process that could be approached strictly computationally is the need for inspectability, explainability, and quality control. That is, we designed the combination-generation algorithm so that it produces useful interim results that can be both pruned and scored using linguistic heuristics. This cuts down on the generation of nonsensical but formally possible combinations, and it is a better approximation of what human processing might be like.

Each combination that makes it through the process of SynMapping without being culled out is called a SynMap. We illustrate its contents using the sentence *The kid dribbled the ball*, one of whose SynMaps is shown in table 4.5.

Column 1

- The first word of the sentence is *the*, so it has the index The-0 (word numbering starts with 0).
- It is analyzed in this SynMap using the lexical sense the-art1.
- The-art1 takes an argument (a noun) that is listed as var1 its syn-struc.
- In this SynMap, that argument is filled by kid-1—the word *kid* in the second position of the sentence.

Column 2

- The second word in the sentence is *kid*, so it has the index kid-1.
- It is analyzed in this SynMap using the lexical sense kid-n1.
- Kid-n1 is a plain noun, so it takes no arguments.

Table 4.5
One SynMap for *The kid dribbled the ball*.

The-0 (the-art1)	kid-1 (kid-n1)	dribbled-2 (dribble-v2)	the-3 (the-art1)	ball-4 (ball-n1)
var0 = The-0	var0 = kid-1	var0 = dribbled-2	var0 = the-3	var0 = ball-4
var1 = kid-1		var1 = kid-1 (subj)	var1 = ball-4	
		var2 = ball-4 (dobj)		

Column 3

- The third word in the sentence is *dribbled*, so it has the index dribbled-2.
- It is analyzed in this SynMap using the lexical sense dribble-v2.
- Dribble-v2 is a transitive verb, so it takes two arguments, which are listed in its syn-struc as var1 and var2.
- In this SynMap, var1 (the subject) is filled by kid-1, and var2 (the direct object) is filled by ball-4.

Columns 4 and 5 are analogous to 1 and 2. The SynMapping process will produce another SynMap for this input, which will be identical to this one except that *kid* will be analyzed using the sense kid-n2, which refers to a baby goat. By contrast, the intransitive meaning of *dribble*—as in *The baby is dribbling*—will not be used in any SynMaps because it cannot account for the direct object in the input.

To recap, SynMapping prepares candidate solutions for the semantic analyzer by establishing candidate dependency structures and pruning out word senses that are incompatible with the syntactic needs of the input.

Basic Semantics   Basic Semantics is responsible for disambiguating the words in a sentence and establishing the semantic dependency structure. Consider examples (4.1) and (4.2):

(4.1) Eleanor cooked hot dogs.
(4.2) Eleanor cooked the books.

And consider just two senses of each of the main constituents:

- *Cook* can refer to preparing food or falsifying something.
- *Hot dogs* can refer to frankfurters or overly warm canine pets.
- *Books* can refer to large written works or financial records.

(4.1) means that Eleanor prepared a food otherwise known as frankfurters, and (4.2) means that she falsified financial records. The other combinations of meanings do not work based on the semantic constraints and preferences listed in the lexicon and ontology. Namely, DOGS are not valid THEMES of PREPARE-FOOD or FALSIFY; BOOK-DOCUMENTS and FINANCIAL-RECORDS are not valid THEMES of PREPARE-FOOD; and FINANCIAL-RECORDS are preferred THEMES of FALSIFY based on a lexical sense for the idiomatic construction *cook the books*.

For examples as short and clearcut as this, Basic Semantics might seem trivial. However, given that most sentences are much more complex and that words can have many more than two senses, the set of candidate combinations can quickly become impressively large.[9] Our computational model of Basic Semantics seeks to emulate human processing by having the agent reject and prefer exactly the same candidate analyses as people would, and for the same reasons—with *reasons* being interpreted functionally since we are working with software and hardware, not wetware.

We present the model of Basic Semantics in the online appendix for inspection by developers. For general readers, the main point is that the combinations of semantic representations are treated in a completely generalized way. That is, the Basic Semantics algorithm contains no special cases, which protects the process from becoming unwieldy as we work toward enabling the agent to understand texts in any domain and of any complexity. When lexical senses require procedural semantic functions, those function calls are recorded as metadata in the TMRs that are the output of Basic Semantics. For example, the pronoun-heavy input *She saw him* results in the following preliminary TMR, whose metadata is shown in gray:

INVOLUNTARY-VISUAL-EVENT-1
    EXPERIENCER   HUMAN-1
    THEME   HUMAN-2
    TIME   *< find-anchor-time*
HUMAN-1
    HAS-GENDER   female
    COREF   resolve-coref-she
HUMAN-2
    HAS-GENDER   male
    COREF   resolve-coref-he

This metadata tells the agent that, during the next stage of processing, it must carry out the procedural-semantic reasoning to identify which individuals in the context are being referred to.

Lexically Triggered Procedural Semantics Lexically Triggered Procedural Semantics covers a lot of territory, including the whole spectrum of coreference issues, context-sensitive adverbs like *respectively* and *very*, and more. Chapter 5 is wholly devoted to coreference and provides ample details of the contribution of Lexically Triggered Procedural Semantics to construction semantics overall.

To conclude this section on construction semantics, let us recap. Construction semantics formally covers the first four stages of language understanding,

but the first one, Basic Syntax, is not cognitively modeled since it is outsourced to data-driven processors. Construction semantics is a noteworthy
microtheory because (a) language-endowed agents must be adept at treating linguistic constructions, and (b) this microtheory illustrates how LEIA
development makes theoretical contributions to cognitive science—in this
case, by extending the purview and implementability of the human-oriented
theory of construction grammar. The following are among the mechanisms
for modeling construction semantics.

- A WMLexicon is dynamically generated and used to process language
  inputs.
- Senses in the WMLexicon are dynamically modified to account for
  transformations.
- Transformations are treated as combined syntactic and semantic
  operations.
- Combinations of transformations are prepared for both explicitly, by
  recording complex transformations, and dynamically, by classifying
  transformations and imposing a class-based ordering on their operation.
- Transformed lexical senses are used as input for SynMapping, which generates candidate solutions based on the syntactic compatibility of word
  senses.
- Both OntoSyntax and Basic Semantics are modeled as wholly generic
  processes. All special cases are treated as part of Lexically Triggered Procedural Semantics.

The modules of construction semantics are followed by two additional
stages of language understanding, Extended Semantics and Situational
Reasoning, which do not rely on heuristic evidence from syntax. By the
time the agent reaches these stages, it is operating exclusively over meaning representations and is incorporating into the language understanding process general reasoning about the world, the situation, its task, and
so on.[10] For illustration, we will give just one example of a phenomenon
treated in Extended Semantics. It is noteworthy because it resonates with
the recent advances in the treatment of complex properties that were
described in section 3.2.1.

When an agent interprets language inputs, how comprehensive should
those interpretations be with respect to complex properties? Should the
agent only record the meaning of what is stated overtly, or should it make

available inferences as well? According to our current model, the agent proceeds as follows:

1. If the TMR contains a state-oriented complex property—for example, X (HAS-SPOUSE Y)—that is the EFFECT of an event—for example, MARRY (AGENT X, Y)—then the agent explicitly stores both the state and the event that gave rise to it.

2. If the TMR contains an EVENT that results in a state—for example, the EFFECT of MARRY (AGENT X, Y) is X (HAS-SPOUSE Y) and Y (HAS-SPOUSE X)—then, as long as the event actually took place (it was not negated, hypothesized about, and so on), the agent explicitly stores both the event and its resulting state.

3. If the TMR contains a complex property that generalizes over repeating events—for example, DIFFICULTY-SWALLOWING .3—then the agent infers and stores an associated set of events—for example, SET (MEMBER-TYPE SWALLOW) (QUANT .8) (DIFFICULTY-ATTRIBUTE .3).

4. If the TMR contains an event—for example, SWALLOW (DIFFICULTY-ATTRIBUTE .3)—which, when repeated, is associated with a complex property—for example, DIFFICULTY-SWALLOWING .3—then the agent does *not* infer the complex property. Why? Because of the murky criteria for "repeating event." For example, if a person has five instances of difficulty swallowing over a ten-year period—each time having wolfed down a huge chunk of steak—he or she does not have a positive value for the symptom DIFFICULTY-SWALLOWING. For this property to be medically relevant, the event instances have to be within a short, recent timeframe. By contrast, IS-REPEAT-CRIMINAL is a binary literal attribute in the ontology and the second instance of committing a crime is enough to set the value to "yes." In short, the agent stores repetition-oriented inferences only if they are confident and straightforward.

5. If the TMR contains a complex property (SIZE) that generalizes over other properties (LENGTH, WIDTH, DEPTH, and so on), the agent does not, by default, attempt to concretize the value. So, if it encounters *a large branch*, it will generate the TMR TREE-BRANCH (SIZE .8) and will do no further inferencing.

6. If the TMR contains an abstract value of a scalar property (WEIGHT), the agent does not, by default, attempt to concretize the value. So, if it encounters *a light pocketbook*, it will generate the TMR PURSE (WEIGHT .2) and will do no further calculation.

7. If the TMR contains a complex property whose definition includes semantic expansions (e.g., HAS-TYPICAL-SYMPTOM), the agent does not record the expansions.

8. If a TMR matches the semantic expansion for a given property, the agent infers that property since that is the reason for recording semantic expansions to begin with. For example, if a teaching physician says *Patients with achalasia complain of difficulty swallowing*, the agent infers and records ACHALASIA (HAS-TYPICAL-SYMPTOM DYSPHAGIA).

9. If a qualitative property (BESIDE) can have quantitative implications (the maximum DISTANCE that can be between X and Y such that they are still BESIDE each other) the agent does not, by default, do any calculations.

The above directives are defaults that assume that the agent does not have any immediate need to make inferences or calculations to achieve its current goals.

This example of property-based inferencing illustrates that LEIAs are designed to extract the full meaning that a human would from language utterances. However, whereas humans don't pay attention to which aspects of language meaning are explicit and which are implicatures, developers of LEIAs need to explicitly model this. Such interaction between language understanding and overall agent reasoning underscores the holistic nature of LEIA modeling and is illustrated by the examples of dialog, learning, and explaining in the chapters to come.

4.3 Language Generation

What *is* natural language generation? Under the broadest definition, it is any automatic process that outputs sentences of natural language. However, the nature of that process makes all the difference in understanding the status and utility of sentences that are generated.[11]

At the time of writing, society at large is most familiar with language generation by large language models (LLMs). In response to a textual prompt, they generate text that is grammatically correct and, in many cases, sensibly responds to the prompt. However, their underlying computational methods and uses in no way resemble language generation by LEIAs.[12]

LEIAs generate language when they have a particular message they want to communicate and they have decided to communicate it using language. Figure 4.2 indicates in boldface which modules from figure 4.1 contribute to language generation.

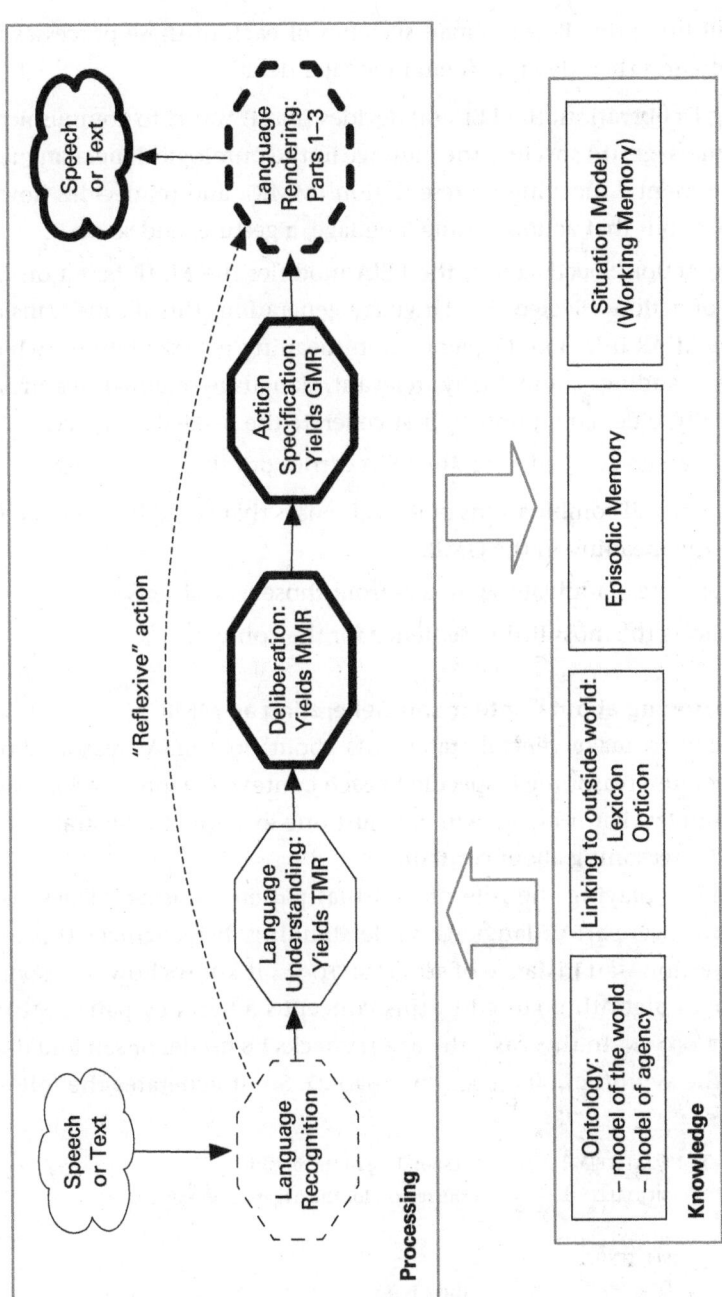

Figure 4.2
A repeat of figure 4.1 with the architecture modules involving language generation indicated by boldface.

We will first provide thumbnail sketches of each of these processes for orientation and then describe them in greater detail.

1. During Deliberation, the LEIA (a) decides that it wants to communicate some message, (b) specifies the message in the ontological metalanguage using a mental meaning representation (MMR), and (c) decides how it will carry out that action—using language, a gesture, and so on.

2. During Action Specification, the LEIA modifies the MMR based on the mode of action selected. For language generation, this means translating the MMR into a GMR (generation meaning representation), which involves adding contextually relevant, language-oriented features—essentially, clues about how to best generate the needed sentence.

3. During Language Rendering, the LEIA carries out three processes:

   a. It creates all combinations of lexical senses that could be used to convey the meaning in the GMR.

   b. It generates candidate sentences from those lexical senses.

   c. It selects the most fitting sentence for the context.

4.3.1   Reasoning about Content and Generating an MMR

It is difficult to make general statements about how LEIAs reason about content because reasoning is specific to each context. Chapters 6–8 provide many examples. For now, we will give just one example to illustrate what we mean by reasoning about content.

Say a LEIA playing the role of a virtual patient is asked "Does your throat hurt?" As part of language understanding, it recognizes this as a yes-no question—an instance of REQUEST-INFO-YN. It knows how to respond thanks to an algorithm stored in this concept's adjacency pair: RESPOND-TO-REQUEST-INFO-YN. In this case, the agent checks its model of self and does not find the symptom PAIN (LOCATION THROAT). So, it generates the following MMR:

ANSWER-YN-QU-NEGATIVELY-1		; responding in the negative
CAUSED-BY	MODALITY-1	; the reason for the negative response is
MODALITY-1		
TYPE	EPISTEMIC	
VALUE	0	; there is no
SCOPE	PAIN-1	; pain

PAIN-1

    LOCATION   THROAT-1   ; in the throat

    TIME   *anchor-time*   ; now

THROAT-1   ; the throat in question is

    PART-OF-OBJECT   LEIA-1   ; part of the LEIA

This MMR is headed by an instance of the concept ANSWER-YN-QU-NEGATIVELY, which means that the agent is planning to respond negatively to this yes-no question. The trace of why it is generating this response is recorded in the CAUSED-BY slot. In plain English, the agent is responding negatively because it is not experiencing pain in its throat. All of the processing needed to generate this MMR is part of reasoning about content. The output MMR serves as input to language generation.

For the practical work of modeling generic, broad-coverage language generation capabilities, it is useful to have a lot of MMRs to practice on. The problem is that, at the current state of the art, LEIAs do not reason in enough ways about enough different things to provide a hefty inventory. But it turns out that this is not a problem because we can use the LEIA's own natural language understanding system to create MMRs to support work on generation. The process is shown in figure 4.3.

Developers select texts whose meanings they want to serve as the starting point for language generation. They send those texts through the LEIA's

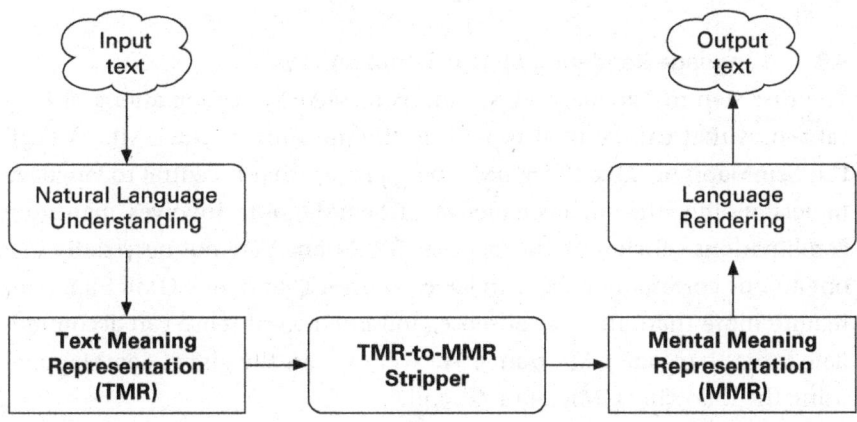

Figure 4.3
How LEIAs can generate MMRs from TMRs to foster work on Language Rendering.

natural language understanding system, resulting in TMRs. Next, a special program—the TMR-to-MMR Stripper—removes all traces of how the meaning was originally conveyed in language, such as the indication of which words and lexical senses gave rise to each of the TMR frames. The result of stripping is meaning representations that are equivalent to the output of the agent's reasoning module; in other words, functionally, they are MMRs. Using automatic language understanding to foster work on language generation is a useful development methodology that is unrelated to configuring end systems.

4.3.2 Action Specification: Converting an MMR into a GMR

During Action Specification, the agent converts MMRs into GMRs. This involves using the dialog history and other aspects of context to prefer certain ways of conveying the given meaning. For example, if the agent decides to respond negatively to a yes-no question (ANSWER-YN-QU-NEGATIVELY), it needs to also decide (a) the stylistic register of the response (neutral, polite, rude, and so on); (b) whether or not it will explain its answer; and, if it chooses to explain, (c) which details the explanation will include. The results of all of these decisions are recorded in the GMR. As with any aspect of agent functioning, there is a big distance between what is minimally needed to configure a useful system and what could be incorporated into a system approaching humanlike capabilities.

4.3.3 Language Rendering Step 1: SemMapping

The first step in Language Rendering is to identify combinations of lexical senses that can be used to convey the meaning in the GMR. We call this SemMapping. Like the SynMapping process that is central to language understanding, the main challenge of SemMapping involves managing combinations—including the fact that GMR frames are not necessarily in a one-to-one correspondence with lexical senses. That is, one GMR frame can require more than one lexical sense, and one lexical sense can accommodate more than one GMR frame. We will explain the gist of SemMapping using the following GMR as an example:

EAT-1
    AGENT   DOCTOR-1
    THEME   CUPCAKE-1
    TIME   < *generation.TIME*

CUPCAKE-1

   SIZE   0.7

One of the candidate SemMaps is presented next. The annotations following semi-colons reflect the fact that the SemMapping algorithm breaks down the GMR into triples and then attempts to satisfy the requirements of argument-taking heads in terms of those triples.

eat-v1

   EAT   ; matches EAT-1 (INSTANCE-OF EAT)

   AGENT $var1   ; matches EAT-1 (AGENT DOCTOR-1)

   THEME $var2   ; matches EAT-1 (THEME CUPCAKE-1)

doctor-n1

   DOCTOR   ; matches DOCTOR-1 (INSTANCE-OF DOCTOR)

cupcake-n1

   CUPCAKE   ; matches CUPCAKE-1 (INSTANCE-OF CUPCAKE)

large-adj1

   $var1   ; no triple-matching required

   SIZE .7   ; matches CUPCAKE-1 (SIZE 0.7)

This structure says:

- The meaning of the GMR can be expressed by a sentence that is generated using the lexical senses eat-v1, doctor-n1, cupcake-n1, and large-adj1.

- The AGENT and THEME slots of EAT are filled by DOCTOR and CUPCAKE, respectively; so, the subject and direct object slots of *eat* are filled by *doctor* and *cupcake*, respectively.

- The SIZE attribute, with a value of 0.7, modifies CUPCAKE; so, *large* modifies *cupcake*.

Several aspects of SemMapping deserve note.

1. The process must cast a wide net in the lexicon in order to capture the many different ways that meanings can be expressed. Even our simple example could be rendered in various ways, as by using the noun *doctor*, *physician*, or *MD*; by using the verb form *ate*, *had*, or *consumed*; by describing the cupcake as *big*, *good-sized*, or *large*; and by formulating the cupcake's noun phrase as *a big/good-sized/large cupcake* versus *a cupcake that was big/good-sized/large*. Moreover, when there are multiple EVENTS in a GMR, they might need to be rendered as verbal heads of clauses or as case role fillers. For example, the GMR frame

POSTPONE-1
- AGENT   TEACHER-1
- THEME   EXAMINATION-1
- TIME   < *generation.TIME*

could be rendered as *The teacher postponed the test, The test was postponed by the teacher, the teacher's postponement of the test, the postponement of the test by the teacher,* and so on. Which one is selected depends on the rest of the GMR content as well as the preceding context.

2. Stylistic features, such as politeness and formality, narrow down the candidates in some cases. The contextually appropriate values of politeness and formality must be established during GMR creation. If none are specified, then the agent can use lexical senses that have either neutral or non-specified values for these features.

3. The SemMapper assumes that the concepts used in the GMR are at the level of ontological precision that the agent wants to generate. For example, if a GMR contains an instance of the concept BUILDING, then the agent will search the lexicon for realizations of BUILDING, not BANK (a hyponym) or PHYSICAL-OBJECT (a hypernym). This generalization does not apply to pronoun use: pronouns are always added to the candidate space in case they are needed.[13] It is actually possible for entities within a chain of coreference to be referred to at different levels of specificity—for example, *beagle/dog, senator/politician,* and *bank/building*—so this capability does eventually need to be modeled but it is not a first priority.

4. SemMapping does not need to address certain details that are taken care of by the data-driven tool that we use to generate sentences from feature-decorated inventories of words.

In sum, during SemMapping the agent must create all semantically valid combinations of lexical senses that might be used to render the GMR's meaning.

4.3.4 Language Rendering Step 2: Generating Sentences from SemMaps

To generate sentences from SemMaps, the agent uses the software package called SimpleNLG (Gatt & Reiter, 2009).[14] Without delving into its particulars, some of the features and values that the agent can provide to SimpleNLG are:

- the set of words that will comprise the sentence and their preliminary ordering—which can be overridden, for example, if the verbal feature "passive" is selected;
- the part of speech of each of the words;
- their syntactic function: for example, for nouns—subject, direct object, and so on; for verbs—main verb, auxiliary verb;
- features of verbs: for example, their form—infinitive, imperative, and so on; their tense—present, past, future;
- features of nouns: for example, number—singular, plural; whether or not they are possessive;
- features of clauses: for example, interrogative type—yes/no; voice—active, passive; and
- components of to-be sentential constituents: for example, a prepositional phrase might need to be formed from the preposition *to* and the noun phrase *a nice cat.*

Among SimpleNLG's capabilities are generating the correct morphological forms of words based on features, generating the passive voice, and generating various forms of possessives. The output of this second step of language rendering is a set of sentences, any of which might be the best choice in the given context.

4.3.5 Language Rendering Step 3: Selecting the Best Sentence

To select the contextually best sentence, the LEIAs use an LLM-based sentence selector.[15] The beauty of using an LLM for this purpose is that all of the candidates passed to it are guaranteed to be semantically correct, so it cannot introduce an actual error. In the worst case, it will select an option that is pragmatically suboptimal. However, testing so far suggests that its performance is actually quite good. Figure 4.4 illustrates LLM-based sentence selection using a rich set of options that were manually created to highlight the LLM's capabilities.

- Given the prompt *Fred jumped off the staircase onto his grandmother's couch,*
- and given the following set of candidate sentences to continue the text

  - Fred heard the springs snap and he realized that he had broken the couch.

Text Selection Test: GPT APIs for LEIA

Enter Paragraph to complete

Enter Paragraph

Enter options separated by ','.

Enter Options

Stochasticity/ Temperature

0.60

Choose a training file:

Zero Shot ›

Predict

Show

<u>Best Fit</u>
- He heard the springs snap and realized he had broken it.

<u>Paragraph:</u>
Fred jumped off the staircase onto his grandmother's couch

<u>Options</u>
- Fred heard the springs snap and he realized that he had broken the couch.
- He heard the springs snap and Fred realized he had broken it.
- He heard the springs snap and realized he had broken it.
- He heard the springs snap and realized that it had been broken by him.
- He heard the springs of the couch snap and he realized that he had broken his grandmother's couch.
- He heard the springs of her couch snap and he realized that he broke her couch.
- He heard the springs snap and realized that it broke.

Figure 4.4

An example of the LLM-based text selector developed in our lab to select among candidate sentences during language generation. The right-hand side shows the results of a system run. The left-hand side is empty because, after an input is processed, the text fields are erased, preparing for the next input.

- He heard the springs snap and Fred realized he had broken it.
- He heard the springs snap and realized he had broken it.
- He heard the springs snap and realized that it had been broken by him.
- He heard the springs of the couch snap and he realized that he had broken his grandmother's couch.
- He heard the springs of her couch snap and he realized that he broke her couch.
- He heard the springs snap and realized that it broke.

- the LLM-based sentence selector chooses the best option: *He heard the springs snap and realized he had broken it.*

Considering the strong performance of the LLM on this task, we see no practical reason to work on developing a symbolic computational approach to selecting among sentence candidates. As a scientific enterprise, a symbolic approach could be of interest since there *are* explainable reasons why people choose one or another word, expression, and syntactic structure, and understanding them is part of understanding how human language works.[16] However, given the long list of LEIA functionalities that cannot be outsourced to data-driven tools, explanatory modeling of this type is not on agenda.

4.4 Comparisons with Other Linguistic Theories

The theory of language processing underlying LEIAs is called Ontological Semantics (Nirenburg & Raskin, 2004). As a theory of computational linguistics, it cares about computability by artificial systems while still being inspired by hypotheses about human language processing. By contrast, theoretical linguistics focuses squarely on human language processing; theoreticians are not concerned about the demands of computer implementations. Nevertheless, two human-oriented linguistic theories have contributed to our thinking: generative grammar and construction grammar.

Generative grammar Generative grammar seeks to model the hypothesized Universal Grammar in human cognition (Chomsky, 1965). Work on the theory involves developing, testing, and modifying the model based on crosslinguistic evidence. Studies within the generative paradigm tend to involve relatively little descriptive detail since it is grammaticality in principle, rather than usage conditions, that is the focus of attention. Hypotheses can be overturned by counterevidence as the theory develops.

Generative grammar is not computational and does not account for how words are selected, how their morphological forms are produced, how they are combined into sentences, or what they mean. Essentially, fully formed sentences are tested for their grammaticality. However, early versions of the theory (prior to the Minimalist Program; Chomsky, 1995) gave rise to a particular understanding of syntax that has influenced both the history of natural language processing—especially with respect to syntactic parsing—and the way that we approach syntactic analysis. For example:

1. The lexicon is considered a listing of both non-argument-taking lexemes and the basic forms of argument-taking lexemes—for example, the active voice of verbs.

2. Lexical structures are projected into the syntax, giving rise to well-formed syntactic structures represented as trees consisting of phrase-level constituents.

3. There are syntactic operations, such as movement and ellipsis, that change the basic structures stored in the lexicon into the many realizations of them permitted in natural language. These transformations account for such things as the passive voice of verbs (*Lou was seen by Jill*), questions (*Who did Jill see?*, *Who was it that Jill saw?*), and verbs used as complements of other verbs (*Lou couldn't see Jill*).

4. Different languages can be treated using similar methods if one models linguistic phenomena in terms of parameters and values.

Construction grammar   Construction grammar is a theory of human language processing that focuses on the form-to-meaning mapping of linguistic structures. So, unlike generative grammar, construction grammar *is* interested in semantics but, due to its human orientation, it only lightly informs computational modeling. A handful of juxtapositions will explain why.

1. Construction grammar is actually not a single theory; it is a diverse set of theoretical and experimental approaches with different foci (cf. point 5 below). For example, Adele Goldberg asks how learners acquire generalizations about constructions and how people can use language creatively while still avoiding expressions that sound like mistakes (Goldberg, 2006, p. 11; Goldberg, 2019, p. 3); Joan Bybee investigates how construction frequency interacts with memory (Bybee 2013, p. 49); and Martin Kay asks what formal diagnostics make a construction a construction and proposes strict rule-in and rule-out conditions for class membership (Kay, 2013).

2. Although construction grammar involves syntax-to-semantics mappings, the syntactic side is far better developed. For this reason, the name construction *grammar* is appropriate. What construction grammarians call semantic analysis is typically an English paraphrase of the given expression.[17]

3. Both the syntactic and semantic sides of constructions tend to be insufficiently specified to be clear to anyone but native speakers.[18] Two examples:

    a. Hoffmann and Trousdale (2013, 2) present the resultative construction as *X V Y Z* and say that it means *X causes Y to become Z by V-ing*, as in *She rocks the baby to sleep*. However, syntactically, X, Y, and Z cannot be just any categories, they have to be specific types of categories; and semantically, *she* is not causing the baby to "become sleep"—instead, she is causing the baby to experience the beginning phase of sleeping.

    b. Goldberg (2013, 17) describes the PN [Preposition BareNoun] construction—for example, *(be) in prison, (go) to school*—as representing a prototypical activity. This is true, but it is too vague to either serve as a sufficient semantic analysis for any particular PN instance (*being in jail* vs. *being on vacation*) or to predict which PN constructions belong to the language. For example, in English one cannot *be in kitchen* even though one eats there as a prototypical activity. Since very few PN constructions work in English, it makes more sense for agent modeling to record them explicitly in the lexicon rather than have a generative rule that will almost always fail because the rule-in cases are so few and idiosyncratic.

4. Construction grammarians reject syntactic transformations and empty categories, instead analyzing sentences by reference to static constructions. For example, Goldberg (2013, 28) says that the sentence *What did Mina buy Mel?* uses the following constructions: Ditransitive construction, Nonsubject Question construction, Subject-Auxiliary Inversion construction, VP construction, NP construction, Indefinite Determiner construction, and Lexical constructions for *Mina, buy, Mel, what, do*. However, construction grammarians do not, to our knowledge, propose algorithms to explain how constructions combine during language analysis and generation.

5. Although most work on construction grammar is not computational, so-called *embodied construction grammar* is. However, the emphasis, at least

so far, has been on computer science angles of embodied language processing, not on linguistic content. Contributions focus on architecture, infrastructure, computation, tools, and narrowly defined applications in which a small number of examples are shown to work.[19] Bergen and Chang (2013) say, "The [ECG] formalism itself is . . . not a linguistic theory per se; rather, it is a theory of what conceptual distinctions are necessary and sufficient to account for language phenomena" (p. 170). As we interpret it, this means that this paradigm provides an infrastructure for content, should that content become available. The needed content, of course, would have to include both knowledge and algorithms. This returns us to one of the main emphases of LEIA modeling: its content-centricity.

To fulfill the more rigorous demands of computation, our microtheory of construction semantics must:

- commit to strict formalisms of syntactic and semantic description;
- use an unambiguous, ontologically grounded metalanguage for expressing meaning;
- operationalize the processing of constructions during both language understanding and language generation; and
- move beyond select examples toward broad-scale coverage of language.

5   The Trajectory of Microtheory Development: The Example of Coreference

Material presented in this chapter illustrates the type of work involved in building microtheories, which are at the center of the theory-model-system trichotomy that scaffolds LEIA development (cf. section 2.4). Microtheories are grounded in theories but must be implementable in state-of-the-art systems. They are expected to evolve over time, starting with phenomena that are either the easiest to treat (simpler-first modeling) or the most urgent for a given application. They are designed to allow for the efficient incorporation of future enhancements.

This chapter illustrates microtheory evolution using the example of coreference, which is a complex linguistic phenomenon that has remained challenging for the natural language processing community despite a half century's research efforts. Whereas past work has led to advances in coreference-oriented *engineering*, it has made few inroads into solving difficult problems of *content*.[1] In the spirit of content-centric cognitive modeling, content is our primary concern.

We have written extensively about coreference in the past and encourage readers unacquainted with this phenomenon to consult the example-rich treatment in McShane and Nirenburg (2021).[2] For readers less interested in linguistic details, the following are the key takeaways from this chapter:

- Computational cognitive modeling requires far greater descriptive rigor than is needed for human-oriented accounts.
- Related phenomena, such as different kinds of referring expressions, should be modeled similarly, as variations on a theme.
- Graphic representations are useful not only for presenting finished models to the outside world but also for the modeling process itself.

- At any given point in model evolution, some examples will be treatable with high precision while others will not. Agents must be able to independently distinguish between these types, treat them accordingly, and append each treatment with metadata about its quality and confidence.
- Well-constructed models of linguistic phenomena can be ported to other languages given appropriate parameterization. The crosslinguistic applicability of models not only captures the reality that languages are more similar than different; it also increases the cost-effectiveness of using computational cognitive modeling to advance artificial intelligence.

5.1 Introduction

Referring expressions are words and phrases that identify ontological OBJECTS and EVENTS. In agent systems, all referring expressions are represented as the heads of frames in meaning representations (XMRs). Whatever knowledge is learned about those entities is then grounded in agent memory. For example, if a particular person named Nancy is discussed many times across different dialog interactions, the LEIA must link all of the learned information about her to the same anchor in memory, just as a person would do. Grounding to memory is called reference resolution, and LEIAs do it during stage 6 of language understanding, Situational Reasoning (cf. section 4.2.1).

By contrast, *co*reference resolution is linking different mentions of an OBJECT or EVENT, presented within a given context, in a chain of coreference. In (5.1), the expressions in boldface are coreferential: *her* corefers with *she*, *she* corefers with *Nancy*, and all three participate in a chain of coreference.

(5.1) **Nancy** couldn't come hiking because **she** was remodeling **her** house.

In most cases, textual sponsors are *ante*cedents: they precede their coreferents. In such cases, the terms *sponsor* and *antecedent* can be used interchangeably. However, sponsors can also follow their coreferents as *post*cedents. This is illustrated by the right dislocation construction in (5.2), which is used for clarification.

(5.2) I like her, Nancy.

For the past several decades, coreference, like language overall, has been treated within mainstream natural language processing almost exclusively using data-driven methods. But despite significant outlays, the state of the art remains rather primitive. As Poesio, Stuckardt, and Versley (2016) report,

"Basically, we know how to handle the simplest cases of anaphoric refer-ence/coreference, anything beyond that is a challenge" (pp. 490–491). They say that, instead of working on the difficult problems, "we're yet occupied with rather mundane issues such as advanced string-matching heuristics for common and proper nouns, or appropriate lexical resources for elemen-tary strategies, e.g., number-gender matching etc" (p. 488).

The microtheories of coreference reported here advance the state of the art in particular ways, sharing with all microtheories the following features:

1. Individual phenomena—such as personal pronouns, demonstrative pro-nouns, definite descriptions, and verb phrase ellipsis—are treated using their own microtheories, which leverage specific kinds of knowledge and reasoning.

2. The microtheories for related phenomena are similar, reflecting the fact that modeling begins with a big-picture assessment of the problem space.

3. Agents are designed to independently distinguish between cases that they can and cannot treat fully and confidently given their current knowledge.

4. At runtime, agents fully resolve the cases they can and posit underspeci-fied analyses for the others. As with all underspecified outcomes, agents subsequently decide whether or not the resulting analysis is actionable. If not, they decide what to do about it (cf. chapter 2 for actionability).

This chapter focuses on coreference resolution for language inputs. Dif-ferent aspects of the problem are addressed at different points in the agent's six-stage process of language understanding, as illustrated by figure 5.1. Since reference processing is fully integrated into language understanding overall, figure 5.1 is best thought of as a stencil that overlays the language understanding process and occludes everything except for the treatment of reference.

1. During Basic Syntax, an externally developed, data-driven coreference engine provides coreference links for some referring expressions.

   a. For some types of referring expressions—such as reflexive pronouns (e.g., *myself*) and first- and second-person pronouns (*I, we, you*)—the system's results are relatively good. They are not perfect since it can be difficult, for example, to identify live speakers and to track referents

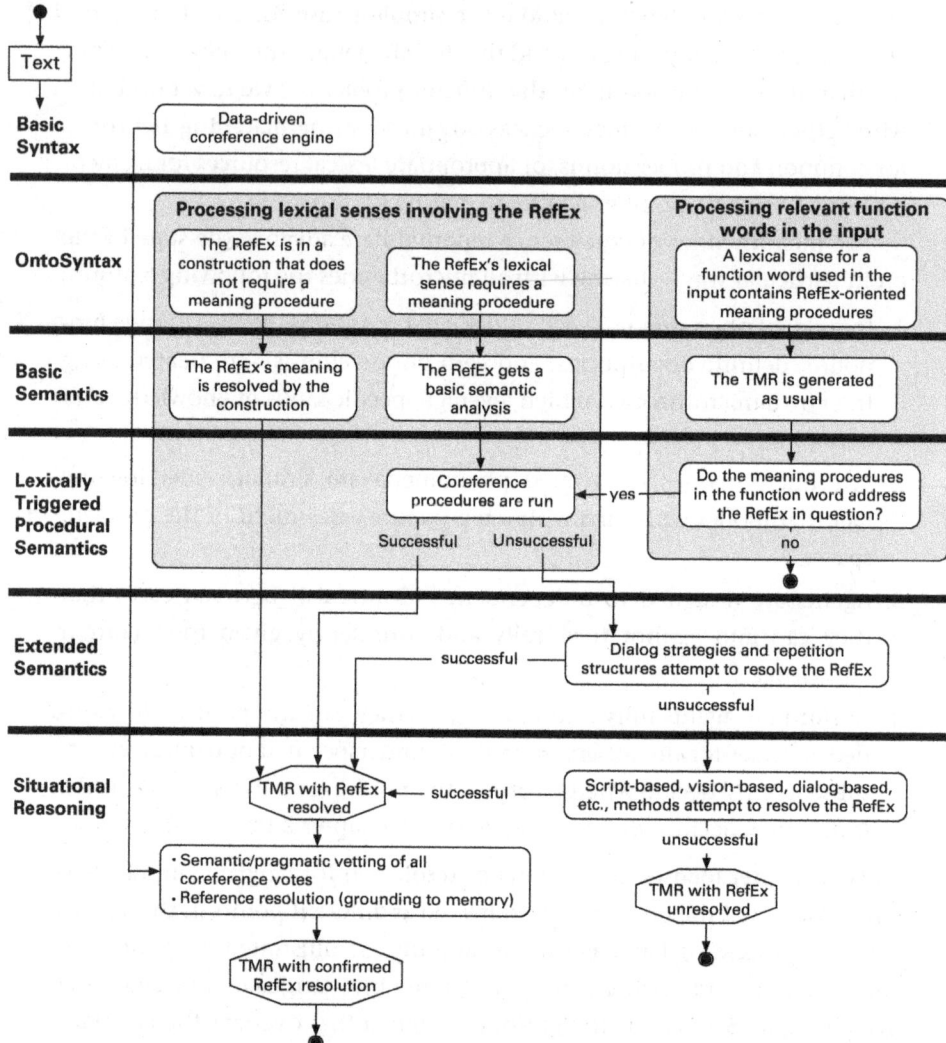

Figure 5.1
Generic algorithm for reference processing. The algorithms for treating specific referring expressions (RefExs) are variations on this theme.

in texts that combine quoted and narrative segments. LEIAs assign such coreference votes high confidence but still attempt to improve upon them using knowledge-based methods.

b. For more difficult kinds of referring expressions—such as third-person pronouns (e.g., *she*) and definite descriptions (e.g., *the driver*)—the data-driven system's results are considerably worse, which is not surprising since semantic and contextual reasoning are often needed. (It is noteworthy that the Winograd Challenge, which was proposed as a way to measure intelligence in computer systems, focuses on resolving third-person pronouns.[3]) Although LEIAs do record the data-driven votes for third-person coreference, they assign them low confidence and attempt to improve upon them using knowledge-based methods.

c. Many referring expressions are not treated at all by the data-driven coreference system, either because they reflect phenomena that are outside of purview (e.g., ellipsis) or because the particular example falls outside of the system's capabilities. For example, the pronouns *it*, *this*, and *that* can corefer with a noun phrase, a proposition, or multiple propositions, but data-driven coreference engines typically identify only noun-phrase sponsors.

2. During OntoSyntax, LEIAs identify all lexical senses that might contribute to resolving the referring expression. This includes

a. lexical senses for the referring expression itself; and

b. lexical senses for function words, like conjunctions, which can suggest coreference relations between entities in the clauses they join; this is further explained in section 5.2.3.

3. During Basic Semantics:

a. LEIAs fully resolve referring expressions in certain kinds of constructions.

b. They posit an underspecified analysis for all other referring expressions: for example, *she* is analyzed as "HUMAN (HAS-GENDER female)," and elided events are detected and analyzed as EVENT.

c. They insert into the meaning representation a call to a procedural semantic routine that, when run at the next stage of processing, will attempt to resolve the coreference.

4. During Lexically Triggered Procedural Semantics, LEIAs run the just-posited procedural semantic routines with two goals:

a. identifying the sponsor; and

b. if needed, making semantic decisions about the relationship between the sponsor and the referring expression, which is not always identity coreference.

5. During Extended Semantics, LEIAs treat as-yet unresolved referring expressions that participate in certain kinds of discourse structures, such as dialog pairs.

6. During Situational Reasoning:

a. For as-yet unresolved referring expressions, LEIAs use any and all evidence available to them—ontological scripts, vision processing, mind-reading of their interlocutor, and so on—to try to identify the sponsor.

b. For all referring expressions that, by this point, have candidate sponsors, LEIAs semantically vet them and ground them to memory, which is the definition of reference resolution, in contrast to *core*ference resolution.

This is the generic model for resolving referring expressions, which serves as a template for the models that treat specific referring expressions.

The first type of coreference we will consider, verb phrase (VP) ellipsis, is arguably the most complex. In fact, it was in solving the problem of VP ellipsis that we understood how best to treat referring expressions overall. This is not merely an autobiographical note; it underscores the fact that cognitive modeling is about trying to understand the unobservable processes of human cognition in a functional way. Linguists have long understood that there are principles of coreference that apply across different kinds of referring expressions. Therefore, it stands to reason that a model that is sufficient to accommodate a particularly difficult case like VP ellipsis could also cover more straightforward cases like personal pronouns.

5.2 Verb Phrase Ellipsis

Verb phrase (VP) ellipsis refers to leaving out a verb phrase when its meaning is clear from the context. In (5.3), the second clause conveys that they might not *go*.

(5.3) They might **go** but they might not __.

Treating VP ellipsis involves three processes: (a) detecting that the verb phrase is missing, (b) identifying its sponsor, and (c) making a battery of semantic decisions about how, precisely, the meaning of the elided VP correlates with the meaning of the sponsor. To understand why all this work is needed, one must understand the nature of this linguistic process.

5.2.1 Linguistic Background and Top-Level Model of VP Ellipsis

Across the world's languages, ellipsis is not merely a stylistic flourish, it is a design feature. Eliding the VP in the second sentence of (5.3) is the most natural way of expressing this idea, and the ellipsis is easy to resolve for three reasons:

1. The sponsor is in the linguistic context. By contrast, in other cases, the sponsor must be recovered from the real-world context. For example, (5.4) could be talking about anything from jumping in the pool to trying a super-spicy curry.

   (5.4) I will ____ if you will ____.

2. The context includes three kinds of **parallelism**, and parallelism fosters ellipsis.[4]

   a. **Syntactic parallelism**: clausal coordination is a kind of structural (syntactic) parallelism in which two clauses of equivalent status, neither being subordinate to the other, are connected using a coordinating conjunction—here, *but*.[5]

   b. **Lexico-semantic parallelism**: this involves repeating the same word in the same meaning—here, *might*.

   c. **Pragmatic parallelism**: it is typical to juxtapose propositions with the same modality but different values, such as *might* and *might not*.

3. The sponsor clause is **simple** since it contains only one main verb. It is not complicated by subordinate, coordinate, or relative clauses whose verbs might, themselves, be the head of the ellipsis sponsor.[6]

   Even if an example has a linguistic sponsor, real-world reasoning can be needed to identify it. In (5.5), in order to select from among the four candidate sponsors—headed by the verbs *call*, *revoke*, *warn*, and *reconsider*—one has to know that if you call on somebody to do something, that person may or may not do it.

   (5.5) The former Massachusetts governor called on United Nations Secretary General Ban Ki-moon to **revoke Ahmadinejad's invitation to the assembly** and warned Washington should reconsider support for the world body if he did not __.·(Gigaword)[7]

Example (5.6) is even more challenging, at least for an agent system, because the sponsor cannot be directly pointed to in the text. Instead, the intended meaning—*have an operation*—is only alluded to by the noun phrase *an operation* three sentences back.

(5.6) "I started with a lot of exercising and then did a lot of physical therapy," the 1992 Olympic 10,000-meter silver medalist said. "I didn't like the idea of an operation. It scared me. I didn't want that. I'm glad I didn't __."(COCA)[8]

In short, VP ellipsis is a microcosm of language overall in that examples range from easily treatable by computer systems to requiring human-level knowledge and reasoning.

As mentioned earlier, LEIAs must be able to identify which examples they can and cannot treat confidently. For the ones they cannot, they must use application-specific reasoning to decide what to do—for example, make do with an underspecified analysis or ask their human collaborator for clarification. Using our model of VP ellipsis, agents automatically divide examples into two categories:[9]

1. Those whose sponsors can be identified using surfacy heuristics, defined as syntactic features supplemented by easily identifiable semantic and pragmatic ones [e.g., (5.3)].

2. Those that require specific world knowledge and associated reasoning to identify the sponsor [e.g., (5.5) and (5.6)]. These are not necessarily outside of an agent's capabilities, they simply rely on particular knowledge being available to it, as opposed to broad-coverage linguistic heuristics.

This binary classification is more than an expedient engineering solution. It is grounded in the observation, long studied in the theory of generative grammar, that the human brain has special mechanisms for processing syntax that are independent of those devoted to semantics. Specifically, generative grammar hypothesizes that there exists a cognitive mechanism, Universal Grammar, that involves the syntactic structure of language, and this mechanism explains how children can achieve the dazzling feat of acquiring the grammar of their native language or languages by around five years old. Both generative grammar and complementary studies in psycholinguistics offer evidence that syntax is at least partially separable from semantics in human language processing. Applied to the case of VP ellipsis, this suggests why, when reading (5.7), we can reconstruct the elided category as *swanged the brole* even though the sentence contains mostly nonsense words.

(5.7) The splot **swanged the brole** and the fot did __ too.

Orienting around syntactic processes to the degree possible is useful for computational cognitive modeling because it allows for near-term progress on a hard problem using simpler-first modeling.

Figure 5.2 shows the top level of the VP ellipsis model.

This model largely parallels the generic model in figure 5.1, but there are some noteworthy differences.

1. During Basic Semantics, nothing is done for VP ellipsis since available data-driven engines do not treat it.

2. During OntoSyntax, LEIAs identify all lexical senses that might contribute to resolving the referring expression. These include:

   a. lexical senses for embedded VP ellipsis constructions, in which the elided VP is embedded within a larger VP that contains its sponsor; for example, *She **ran** as fast as she could __*;

   b. lexical senses for modal words used in constructions that lack a VP complement, which is the diagnostic for VP ellipsis; for example, there is a lexical sense of *didn't* that expects only a subject and no overt VP complement; and

   c. lexical senses for function words, like conjunctions, whose meaning procedures suggest coreference relations between the clauses they join.

3. During Basic Semantics,

   a. LEIAs fully resolve elided VPs in embedded VP constructions like *She ran as fast as she could__*.

   b. For the other elided VPs, they posit an underspecified EVENT analysis and insert into the meaning representation a call to a procedural semantic routine that, when run at the next stage of processing, will attempt to resolve the coreference.

4. During Lexically Triggered Procedural Semantics, LEIAs run the just-posited procedural semantic routines with two goals:

   a. identifying the sponsor; and

   b. making a battery of semantic decisions about how the elided VP corresponds to the sponsor.

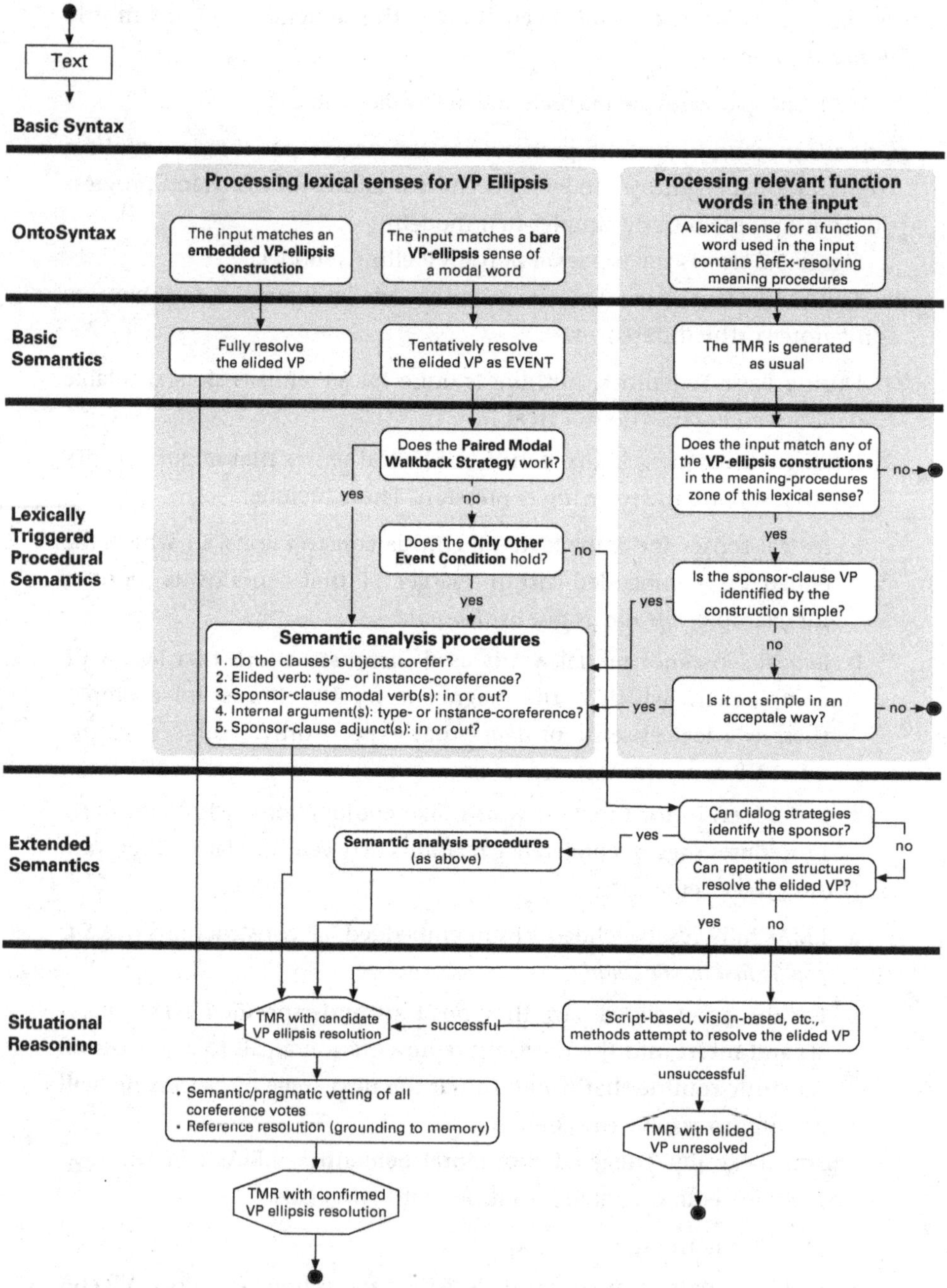

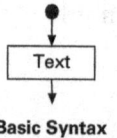

Basic Syntax

	Processing lexical senses for VP Ellipsis		**Processing relevant function words in the input**
OntoSyntax	The input matches an **embedded VP-ellipsis construction**	The input matches a **bare VP-ellipsis** sense of a modal word	A lexical sense for a function word used in the input contains RefEx-resolving meaning procedures
Basic Semantics	Fully resolve the elided VP	Tentatively resolve the elided VP as EVENT	The TMR is generated as usual

Lexically Triggered Procedural Semantics

Does the **Paired Modal Walkback Strategy** work?
— yes
— no

Does the **Only Other Event Condition** hold? — no
— yes

Semantic analysis procedures
1. Do the clauses' subjects corefer?
2. Elided verb: type- or instance-coreference?
3. Sponsor-clause modal verb(s): in or out?
4. Internal argument(s): type- or instance-coreference?
5. Sponsor-clause adjunct(s): in or out?

Does the input match any of the **VP-ellipsis constructions** in the meaning-procedures zone of this lexical sense? — no ➤●

yes

Is the sponsor-clause VP identified by the construction simple? — yes

no

Is it not simple in an acceptable way? — no ➤●

yes

Extended Semantics

Semantic analysis procedures (as above)

Can dialog strategies identify the sponsor? — yes
no

Can repetition structures resolve the elided VP?

yes   no

Situational Reasoning

TMR with candidate VP ellipsis resolution — successful — Script-based, vision-based, etc., methods attempt to resolve the elided VP

unsuccessful

• Semantic/pragmatic vetting of all coreference votes
• Reference resolution (grounding to memory)

TMR with elided VP unresolved

TMR with confirmed VP ellipsis resolution

Figure 5.2
Top level of the model for treating VP ellipsis.

5. During Extended Semantics, LEIAs resolve as-yet unresolved VPs that participate in certain kinds of discourse structures, such as pairs of dialog turns.

6. During Situational Reasoning,

   a. For as-yet unresolved elided VPs, LEIAs use script-based, vision-based, dialog-based, and other evidence to try to identify the sponsor.

   b. For all referring expressions that, by this point, have candidate sponsors, LEIAs semantically vet them and ground them to memory, which is the definition of reference resolution, in contrast to coreference resolution.

A possible outcome is that the LEIA could not resolve the elided VP, in which case it uses the underspecified TMR as input to its reasoning about action.

The role of the right-hand rectangle in figures 5.1 and 5.2 requires further explanation. This portion of the algorithm involves lexical senses for function words, like conjunctions and punctuation marks, which are the heads of syntactic constructions that can suggest coreference links. For example, when *but* joins two clauses, the resulting parallelism suggests that the clauses' subjects, direct objects, and VPs might be coreferential. *Might* is the key word here: such constructions are among many heuristics that must be evaluated in any context.

Figure 5.3 illustrates how the modal verbs and function words collaborate to suggest coreference links.

It is not only syntactic parallelism that can suggest coreference across clauses; semantic parallelism can as well, as illustrated by the *if . . . then* construction in figure 5.4.

To emphasize, constructions like "Clause1 *but* Clause2" and "*If* Clause1 *then* Clause2" do not guarantee coreference across parallel constituents in their clauses. Instead, they are among a cluster of heuristics that, together, inform coreference analysis.

Why do we emphasize that some coreference votes are suggested by conjunctions? Because this guides how and where the associated coreference knowledge is recorded. Whereas it is easy to make generalizations like "feature-matching subjects of conjoined clauses tend to corefer," it is a different matter altogether to transform such generalizations into useful, traceable, and extensible knowledge within a comprehensive agent system. Since it is the parallelism established by *but* in the construction "Clause1

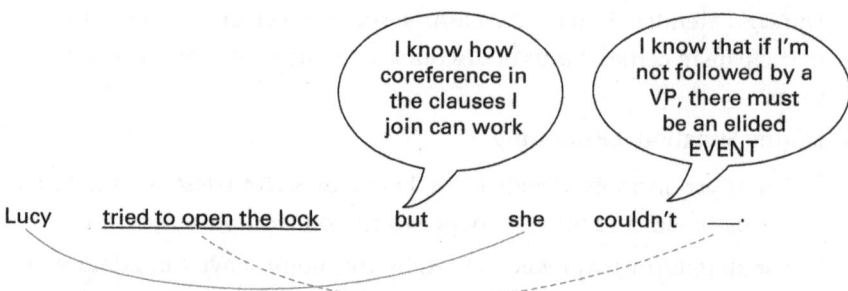

Figure 5.3
The modal verb *couldn't* used in a construction that lacks a VP complement signals the VP ellipsis. The coordinating conjunction *but* suggests potential coreference relations between the subjects and the verb phrases in the coordinated clauses.

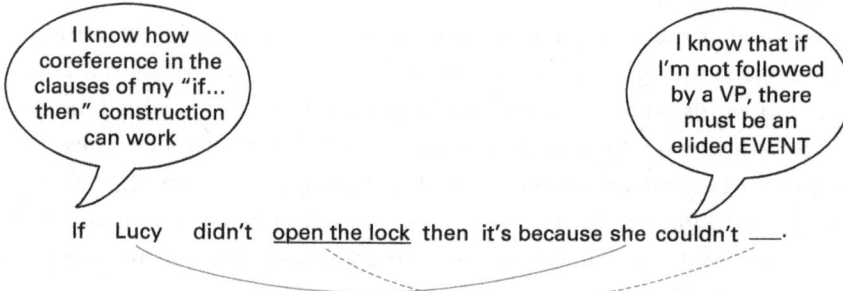

Figure 5.4
The construction *if . . . then* establishes semantic parallelism across the clauses it joins. This suggests potential coreference relations between the parallel constituents—here, the subjects and verb phrases.

but Clause2" that suggests certain coreference relations, the associated coreference rules are stored in the meaning-procedures zone of that sense of *but*. Details of this approach are presented in section 5.2.3.

We have only just scratched the surface of the top-level model of VP ellipsis shown in figure 5.2. The sections below work through the algorithm in greater detail.

5.2.2 Embedded VP Ellipsis Constructions
Beginning with the upper-left box in figure 5.2, embedded VP ellipsis constructions describe inputs in which VP ellipsis is embedded in the sponsor's verb phrase, as illustrated by example (5.8) and figure (5.5).[10]

(5.8) He will give his kids as many gifts as he wants to.

A noteworthy feature of such constructions is that the whole VP cannot be used to resolve the ellipsis because that would lead to endless recursion: *He will give his kids as many gifts as he wants to give his kids as many gifts as he wants to give.* . . . . Next are some additional examples of embedded VP ellipsis constructions, described informally. See this footnote[11] for presentation conventions.

(5.9)   [Subj$_1$ (Modals) $\mathbf{V^1}$ as best (as) Pro$_1$ (possibly) CAN <could>]
They **worked** as best they could __.

(5.10) [Subj$_1$ (Modals) $\mathbf{V^1}$ WHEN Pro$_1$ Modals]
I had to **take my chances** when I could __.

(5.11) [Subj$_1$ (Modals) $\mathbf{V^1}$ WHENEVER <any time> Pro$_1$ Modals]
She **traveled** whenever she wanted to __.

(5.12) [Subj$_1$ (Modals) $\mathbf{V^1}$ HOWEVER <(in) any way that> Pro$_1$ can]
He **earns a living** however he can __.

(5.13) [Subj$_1$ (Modals) $\mathbf{V^1}$ (just) AS Adj/Adv as Pro$_1$ (possibly) can <could>]
I want to **live** as long as I can __.

(5.14) [Subj$_1$ (Modals) $\mathbf{V^1}$ WHERE <anywhere, wherever, any place> Pro$_1$ Modals]
People try to **find pleasure** where they can __.

Deciding which constructions to record, and how to balance literal and variable elements in them, is a good example of the knowledge

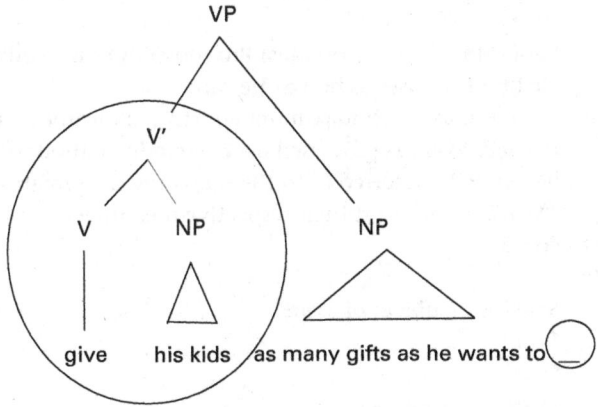

Figure 5.5
A partial syntactic tree for the construction in example (5.8).
The constituent *give his kids* is the sponsor for the elided VP.

Table 5.1
Two constructions accommodate example (5.8). Only the second, more generic, one accommodates (5.15) as well.

(5.8)	He	will	**give**	**his kids**	as many	**gifts**	as	he	wants to	___.
↑	Subj$_1$	(Modals)	V^1	(NP)	as many	N^1	as	Pro$_1$	wants to	___.
↑↓	Subj$_1$	(Modals)	V^1	(NP)	Quant	(N^1)	as	Pro$_1$	Modals	___.
(5.15)	He	will	**run**		as much		as	he	has to	___.

engineering challenge overall. For example, (5.8) can be covered by either of the constructions listed in table 5.1. However, the second, more generic, construction—which treats the quantifier and modal verb as variables and allows for ellipsis of the head noun—also covers examples like *He will run as much as he has to.*

In general, the more narrowly defined the construction, the more precise its stored semantic description can be. However, recording constructions takes time, and more narrowly specified constructions offer less coverage than more broadly specified ones. So, knowledge engineers must find the sweet spot between the generic and the specific.

Examples (5.16) and (5.17) illustrate embedded VP ellipsis constructions along with the lexical senses they require and the TMRs the system produces for them. These analyses follow the leftmost path down figure 5.2.

(5.16) He builds houses as best he can.

The sense of *can* that covers **He builds houses as best he can** is can-mod31.

can-mod31
    definition   Subj (Modals) V^1 as best (as) Pro (possibly) can <could> __.
    example   He builds houses as best as he can.
    comments   Analyzed as "with maximum effort." No meaning procedure is
               needed. $var1 is processed using generic methods, it need not
               be explicitly referred to in the sem-struc (its semantic relation to
               ^$var2 depends on their respective meanings).
    synonyms   could[12]
    syn-struc
      subject $var1   (subject of $var2)
      v^1 $var2
      adv $var3   (root 'as')
      adv $var4   (root 'best')
      adv $var5   (root 'as') (opt+)
      subject $var6   (type pro)

adv	$var7	(root 'possibly') (opt+)
v	$var0	

sem-struc
  MODALITY
    TYPE   EFFORT
    VALUE   1
    SCOPE   ^var2
  $var3,4,5,6,7 null-sem+

The TMR for **He builds houses as best he can:**

MODALITY-1		
TYPE	EFFORT	
VALUE	1	; maximum effort
SCOPE	BUILD-1	; is applied to build
BUILD-1		
AGENT	HUMAN-1	
THEME	PRIVATE-HOME-1	
HUMAN-1		
HAS-GENDER	male	
PRIVATE-HOME-1		
CARDINALITY	>1	; a shorthand for full set notation

  (5.17)   She works when she can.

The sense of *when* that covers **She works when she can** is when-adv6.

when-adv6
  definition   Subj (Modals) V^1 when <whenever, any time that> Pro Modals
          —.
  example   She works when she can.
  comments   In some cases, iteration is implied by the verbal aspect, but
           that will be taken care of by aspect processing that is separate
           from this VP ellipsis construction. $var1 is processed using
           generic methods; it need not be explicitly referred to in the
           sem-struc (its semantic relation to ^$var2 depends on their
           respective meanings).
  synonyms   whenever, any time that
  syn-struc

subject	$var1	(subject of $var2)
v^1	$var2	
adv	$var0	
np	$var3	(type pro)
modals	$var4	

sem-struc
 ^$var2
   TIME   ^$var4.TIME
 ^$var4
   SCOPE   ^$var2
 $var3   null-sem+

The TMR for **She works when she can:**

WORK-ACTIVITY-1
   AGENT   HUMAN-1
   TIME   MODALITY-1.TIME
   TIME   *find-anchor-time*
HUMAN-1
   HAS-GENDER female
MODALITY-1
   TYPE   POTENTIAL
   VALUE   1
   SCOPE   WORK-ACTIVITY-1

Although (5.17) implies multiple instances of working, iteration is not a necessary feature of this construction. For example, *I'll do it when I can* implies a single event instance. Iteration needs to be handled by a microtheory of tense and aspect that is under development.

An important consideration when recording constructions is that they must support not only language analysis but also language generation. If the lexicon were used exclusively for analysis, more bunching of constructions would be possible. For example, can-mod31 could have included the option *well* as a realization of $var4 since *as well as one can* is a common paraphrase. However, with the *well* variant, the second *as* is not optional: one cannot say *as well one can*. For analysis, this would not be a problem since such inputs will likely never occur; but for generation, the agent would have no way of knowing that the second *as* was optional for *best* but not for *well*. Of course, one could introduce additional expressive means into the lexicon to account for such options but that would impose a higher cognitive load on knowledge engineers as well as more computational overhead for programmers. It is faster and easier to create a different lexical sense for the *as well as* variant.

Conceptually, one can think of embedded constructions as the first line of analysis for inputs with VP ellipsis since, if an input matches such a construction, the associated analysis will be the right answer. However, there is no need to create an ordered algorithm to implement this preference.

Instead, all VP ellipsis resolution strategies shown in figure 5.2 are attempted for every input, and each one carries a confidence score, with embedded constructions being scored the highest. So, embedded-construction analyses will always win out without the need to complicate the basic processing flow with an ordered, VP-specific algorithm.

5.2.3   Syntactic Constructions Anchored in Function Words

VP ellipsis is always licensed by a modal word. For clarity of presentation, we will concentrate on modal verbs like *can't* but modal adjective constructions like *isn't able to* and modal noun constructions like *their inability* work similarly.

The lexicon contains a sense of each modal verb that anticipates VP ellipsis in the most minimalistic construction: [Subj ModalVerb]. We call such lexical senses *bare VP ellipsis senses*. Their output-syntax is clause (CL), which means that the listed constituents are understood to be the only major constituents in the given clause. Modifiers are permitted as a matter of course. So, bare VP ellipsis senses can only match inputs with VP ellipsis.

Bare VP ellipsis senses are needed to detect and provisionally resolve VP ellipsis in contexts that are not covered by embedded VP ellipsis constructions—as shown by the top middle box in figure 5.2. As an example, below is the bare-modal sense of *can* indicating potential. (*Can* can also indicate permission, which is covered by a different sense.)

```
can-mod2
    definition          The bare VP ellipsis sense indicating potential.
    example             Doris can.
    output-syntax       CL
    syn-struc
        subject         $var1
        modal           $var0
    sem-struc
      MODALITY
        TYPE            POTENTIAL
        VALUE           1
        SCOPE           EVENT-1
      EVENT-1
        CASE-ROLE-1     ^$var1
    meaning-procedures
      seek-sponsor EVENT-1
      seek-specification CASE-ROLE-1
```

This lexical sense is read as follows:

- The syn-struc says that this construction contains a subject followed by the modal verb *can* and no other major constituents. Importantly, there can be no overt VP following the modal verb.

- The sem-struc says that potential modality with a value of 1 scopes over some EVENT that was elided. It is referred to as EVENT-1, with the index being used for cross-referencing between the sem-struc zone and the meaning-procedures zone.

- The sem-struc also says that the meaning of the subject fills some CASE-ROLE of the as-yet unknown EVENT. Which CASE-ROLE is needed cannot be predetermined because it depends on the meaning of the input. If *Doris can sleep*, then she is the EXPERIENCER-OF SLEEP; if *Doris can cook* then she is the AGENT-OF PREPARE-FOOD; and if *The tractor can pull the car*, then the tractor is the INSTRUMENT-OF PULL. The index on CASE-ROLE is used for cross-referencing between the sem-struc zone and the meaning-procedures zone.

- The meaning-procedures zone contains calls to programs that will attempt to complete the coreference resolution during Lexically Triggered Procedural Semantics. Some resolution procedures come from bare-modal senses, whereas others come from function words used to analyze the input. We have already begun to explain the reason for this division and will now work through the details.

As explained earlier, linguistic phenomena like syntactic and semantic parallelism can suggest coreference relations, as illustrated by the following examples:

(5.18) [clausal coordination]
James₁ **swam** yesterday but he₁ couldn't ___ today.

(5.19) [clausal juxtaposition (parataxis)]
James₁ **swam** yesterday; unfortunately, he₁ couldn't ___ today.

(5.20) [conditional construction]
You₁ can **go** if you₁ want to __

All of these constructions involve VP ellipsis, but they are not uniquely *about* VP ellipsis. They are first and foremost about their own semantics and pragmatics, be it to express contrast, juxtaposition, or a conditional relationship between propositions with specific meanings. Only second-arily do they suggest coreference relations—and not only for VP ellipsis, but for other categories like subjects as well. The question is, how should we record and use the multiple kinds of coreference information conveyed by constructions like these?

One possibility would be to divide each of the above phenomena—contrasts, juxtapositions, and conditionals—into multiple constructions, with each one focusing on a different coreference phenomenon. This is illustrated by figure 5.6 using the example of "Clause1 *but* Clause2" constructions.

Under this approach, every coreference relation suggested by *but* would be treated by a different construction. In essence, this approach says:

- When *but* joins clauses and the second clause contains VP ellipsis, the first clause might contain its sponsor. There's a construction for this.

- When *but* joins clauses and the second clause has a subject pronoun, the first clause might contain its sponsor. There's a construction for this.

- When *but* joins clauses and the second clause has a pronominal direct object, the first clause's direct object might contain its sponsor. There's a construction for this.

Modeling knowledge this way makes some sense: this is conceptually clean and it allows for maximum variability in the components of the

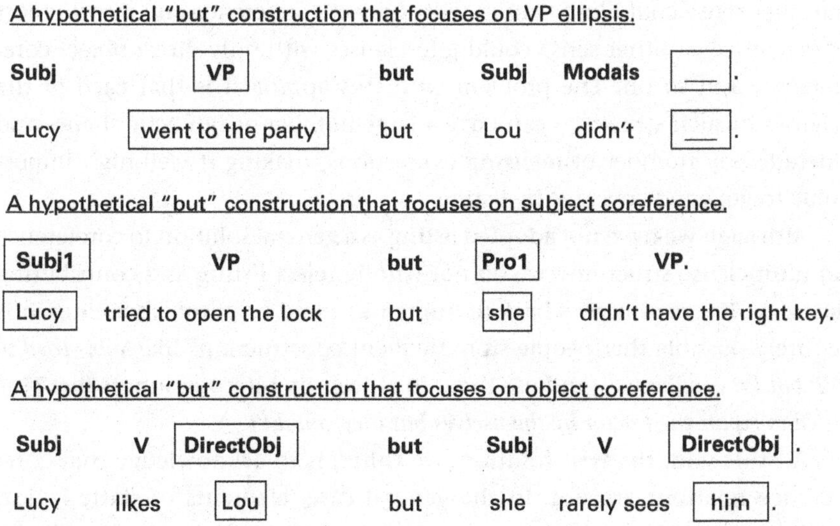

Figure 5.6
Hypothetical *but* constructions that treat one coreference phenomenon each.

construction that are not focused on. For example, the first and third constructions accommodate different subjects, and the second one allows for VP ellipsis. However, implementing this solution would require fundamentally changing the language understanding algorithm used by LEIAs. Namely, agents would need to be permitted to concurrently use multiple senses to analyze a word as long as the analyses unified. For example, when analyzing *Lucy tried to open the lock but she couldn't*, the agent would need to select the first sense of *but* for information about VP ellipsis resolution and the second sense of *but* for information about subject pronoun coreference. These analyses would unify because both would analyze *but* using the relation CONTRAST. Although this modification to the language analysis algorithm might sound reasonable, it would actually introduce a lot of complexity because, instead of selecting one lexical sense to consume each word of input, the agent would need to consider all possible combinations of lexical senses. In short, this would be a bad solution to what is, essentially, an invented problem—"invented" because it arises from an approach to modeling coreference that is in no way necessary.

Another option would be to try to account for all possible combinations of coreference relations using a large set of constructions, each one addressing a particular combination of coreference phenomena. For example, one sense of *but* could join clauses with coreferential subjects and VP ellipsis; another sense could join clauses with subject coreference and direct object coreference; another sense could join clauses with only direct object coreference; and so on. The problem with this approach is that each of the clauses in such structures can have a large number of syntactic shapes and include any number of referring expressions, making it well-nigh impossible to get good coverage by listing.

Although we have not adopted listing as a general solution to coreference in multiclause structures, we do not wholly reject listing as a contributing strategy. The reason goes back to human-inspired cognitive modeling. It is entirely possible that people store frequent constructions like *Subj₁ tried to VP but Pro₁ couldn't* __ and use them to efficiently interpret inputs like *They tried to repair the tractor by themselves but they couldn't* __.

All this said, the best solution, we think, is to acknowledge that coreference relations are not, in the general case, elements of static lexical constructions. Instead, establishing coreference is a procedure, and procedures should attach to whichever lexical senses conceptually spawn

them. We just saw that it is the parallelism established by *but* in the construction "Clause1 *but* Clause2" that suggests various kinds of coreference
relations, so the associated coreference rules are best stored in a sense of
but that covers this construction. The lexical sense for but-conj2 shows
how this works.[13]

but-conj2
definition	'but' conjoining clauses
example	He came but she didn't (come).
comments	The meaning-procedures zone includes calls to coreference procedures based on inter-clause parallelism.

syn-struc
cl	$var1
conj	$var0
cl	$var2

sem-struc
CONTRAST	
DOMAIN	^$var1
RANGE	^$var2

meaning-procedures

TRY: RESOLVE-VP-ELLIPSIS	; Sallie wanted to **come** but she couldn't __.
AND-TRY: RESOLVE-IDENTICAL-SUBJ-PRONOUNS	; **He** came but **he** left early.
ELSE-TRY: RESOLVE-FEATURE-MATCHING-SUBJS	; **My cousin** came but **he** left early.
AND-TRY: RESOLVE-IDENTICAL-PRONOMINAL-DIROBJS	; Ed borrowed **them** but he lost **them**.
ELSE-TRY: RESOLVE-FEATURE-MATCHING-DIROBJS	; Ed borrowed **my keys** but he lost **them**.
Etc.	

The syn-struc of but-conj2 is minimalistic: *but* links two clauses of any
shape and establishes the CONTRAST relation between their meanings. The
rules in the meaning-procedures zone guide coreference processing. They
extend the conditional formalism used in previous examples. AND-TRY
indicates that the function should be evaluated even if the previous one
was run, which prepares for addressing different referring expressions in
turn. ELSE-TRY indicates that the function is to be evaluated only if the
previous one did not run; such functions involve more broadly defined
preconditions than their predecessors. For example, RESOLVE-IDENTICAL-
SUBJ-PRONOUNS tests if consecutive subjects string-match, as in *he . . . he*;
if not, RESOLVE-FEATURE-MATCHING-SUBJS tests if they feature-match, as in *my
cousin . . . he*.[14]

The rules recorded as meaning procedures can be as specific as desired,
including promoting frequent constructions like *Subj₁ tried to VP but Pro₁
couldn't __* to a set of first-priority cases that are evaluated first. Each rule is

indexed so that the metadata in the associated TMR indicates which rule was used to establish each coreference link. The following is an example of a TMR that includes VP ellipsis resolution.[15]

(5.21) David has to **swim** but Dolores wants to ___.

CONTRAST-1		; there is a contrast between
DOMAIN	MODALITY-1	; having to swim
RANGE	MODALITY-2	; and wanting to (swim)
lex-sense	but-conj2	
MODALITY-1		; having to swim
TYPE	OBLIGATIVE	
VALUE	1	
SCOPE	SWIM-1	
TIME	*find-anchor-time*	
lex-sense	have-mod1	
SWIM-1		
AGENT	HUMAN-1	
lex-sense	swim-v1	
HUMAN-1		
HAS-NAME	"David"	
lex-sense	personal-name-n1	
MODALITY-2		; wanting to swim
TYPE	VOLITIVE	
VALUE	1	
SCOPE	SWIM-2	; reflects ellipsis resolution
TIME	*find-anchor-time*	
lex-sense	want-mod2	
SWIM-2		; reflects ellipsis resolution
AGENT	HUMAN-2	
uses-lex-sense	want-mod2	; contributes to the ellipsis resolution
uses-lex-sense	but-conj2	; contributes to the ellipsis resolution
HUMAN-2		
HAS-NAME	"Dolores"	
lex-sense	personal-name-n1	

An important decision point in the VP ellipsis algorithm involves determining whether the example is **simple** enough to be treated without real-world reasoning (in figure 5.2, look at the right-hand gray box in the stage Lexically Triggered Procedural Semantics). An example can either be completely simple, with the sponsor containing only one main verb, or it can be less simple but in ways that do not significantly reduce the reliability of coreference predictions. Examples of the latter are as follows.

(5.22) [The sponsor clause can contain conjoined Vs]
Charlie **came and socialized** but Jane didn't __.

(5.23) [The sponsor clause can contain conjoined VPs as long as each one is simple]
Charlie **went to the park and had an ice cream** but Jane didn't __.

(5.24) [The sponsor clause can contain any number of modals]
Charlie **wanted to try to skateboard** but Jane didn't __.

(5.25) [The elided VP can be subsumed within a matrix clause, shown in italics]
Charlie **washed his car** yesterday but *Jane told us that* he couldn't __ today.

(5.26) [The sponsor clause can have an object-controlled verb]
Charlie wanted Jane to **accompany him to Prague** but she couldn't __.

(5.27) [The input can contain any combination of the above complexities]
Charlie wanted to try to **wash and wax his car** but Jane noticed that he didn't __.

VP ellipsis in coordinate structures establishes a theme for which ellipsis in paratactic and conditional structures are variations. To repeat examples from above:

(5.19) [clausal juxtaposition (parataxis)]
James$_1$ **swam** yesterday; unfortunately, he$_1$ couldn't ___ today.

(5.20) [conditional construction]
You$_1$ can **go** if you$_1$ want to __

Clausal juxtaposition is just like clausal coordination except that the construction is recorded as a sense of a punctuation mark rather than a conjunction, and the meaning of the relation is JUXTAPOSITION (not CONTRAST, as for *but*).[16] As regards conditional structures, they are recorded using senses like *If Clause1(,) (then) Clause2*, and the clauses are semantically linked using the relations PRECONDITION and EFFECT.

An obvious question is, do subordinate conjunctions have similar power to suggest coreference relations across the clauses they join? The reason to wonder, rather than just assume that they do, is that they are not semantically or pragmatically parallel like the previous cases. In very simple examples like (5.28)–(5.31), coreference predictions do seem to work the same; however, this issue requires additional corpus-informed study.

(5.28) Although James wanted to **swim**, he couldn't __.
(5.29) Because Betty had to **swim**, she did __.

(5.30) James **wanted to swim** even though Betty didn't __.
(5.31) James **came and swam** because he could __.

Recap: We have just talked about two classes of constructions that syntactically suggest the sponsor of VP ellipsis: embedded VP ellipsis constructions (e.g., *She ran as fast as she could* __) and syntactic constructions whose parallelism helps to identify the ellipsis sponsor (e.g., *Roberta laughed but Tony didn't* __). So far, we have only talked about syntax with the goal of identifying the textual sponsor of the elided VP. Before turning to the important matter of semantically reconstructing the ellipsis (section 5.2.5), we must consider two additional methods of identifying the textual sponsor for an elided VP.

5.2.4   Other Methods of Identifying Textual Sponsors

Look back at figure 5.2—specifically, at the left-hand, light gray box at the stage of language understanding called Lexically Triggered Procedural Semantics. In includes two resolution strategies called the Paired Modal Walkback Strategy and the Only Other Event Condition, which we will now explain.

The Paired-Modal Walkback Strategy orients around pragmatically typical pairs of modal meanings, what we call *paired modals*, like *can ~ can't* and *tried to ~ couldn't*.

Paired modals are defined as any of the following:

1. The clauses include the same modal value: for example, *can ~ can* (POTENTIAL 1).

2. The clauses include the same modal type but with different values: for example, *can* (POTENTIAL 1) *~ can't* (POTENTIAL 0).

3. The clauses include pragmatically typical, ordered correlations of particular modal values, such as

   a. tried to (EFFORT 1) → couldn't (POTENTIAL 0)

   b. should have (OBLIGATIVE 1) → didn't (EPISTEMIC 0)

   c. wanted to (VOLITIVE 1) → couldn't (POTENTIAL 0)

The idea is that, no matter the syntactic configuration, if the VP complement of the second modal in the pair is elided, it is likely that its sponsor is the complement of the first modal—as long as the modals are quite proximate in the text.[17] The Paired-Modal Walkback strategy catches examples that in some way defy the strict syntactic requirements of the construction-based approach described in the previous section. For example, (5.32) is not treated by syntactic constructions because the context is not simple or acceptably non-simple due to the extra sentence between the sponsor clause and the ellipsis clause.

(5.32) He can't **whistle?** That's odd. I thought he could __.

The nascent TMR for this input after Basic Semantics includes (among others) the following frames, which show the paired modals:

He can't whistle   (I thought) he could

MODALITY-1   MODALITY-2
  TYPE   POTENTIAL   TYPE   POTENTIAL
  VALUE 0   VALUE 1
  SCOPE WHISTLE-EVENT-1   SCOPE EVENT-1

Can't indicates potential modality with a value of 0, whereas *could* indicates potential modality with a value of 1. This pairing suggests, on semantic grounds, that that EVENT-1 corefers with WHISTLE-EVENT-1.

Orienting around pairs of modal meanings extends the reach of this microtheory in two ways. First, it allows for idiosyncratic additional constituents to intervene in what otherwise might be recognized as an ellipsis-predicting syntactic construction. Whereas (5.32) is not simple because of the intervening sentence, (5.33) is not simple because the sponsor clause contains multiple verbs, each of which could, in principle, be the head of the ellipsis sponsor.

(5.33) Sharon won't **come and help you trim the hedges before you leave for vacation?** I thought she would __.

However, because of the modal pair, the ellipsis can still be confidently resolved.

The second benefit to orienting around pairs of modal meanings is that the modality can be expressed in nonverbal ways as well, such as by using a noun like *ability*.

(5.34) Ryan's ability to **singlehandedly repair a chain-link fence** doesn't mean that I can __.

Clearly, there is no syntactic parallelism when the modal meaning in the sponsor clause is conveyed by a noun (*ability*) whereas the modal meaning in the ellipsis clause is conveyed by a verb (*can*). But there *is* semantic parallelism (both *ability* and *can* express potential modality), which the Paired-Modal Walkback Strategy exploits.

The second situation covered in this corner of the VP ellipsis algorithm involves what we call the Only Other Event Condition. If the preceding context contains only one event, then that event might be the sponsor. This can occur, for example, in the opening of a novel:

(5.35) She **quit**. Empty office. Bare shelves. He never thought she would __.

Here, the elided VP and its sponsor are separated by two sentences, but neither of them includes an EVENT that could serve as an ellipsis sponsor. Although the Only Other Event condition is useful, it is not reliable since the actual sponsor might not be an antecedent—it might be a postcedent or it might be recoverable only through the nonlinguistic context. Therefore, this resolution strategy results in a soft hypothesis that, like all hypotheses, will be further evaluated in later stages of processing.

If neither of these semantically oriented resolution strategies identifies a candidate sponsor, then the provisional EVENT, which was the initial reconstruction of the elided VP, remains underspecified, awaiting possible resolution downstream (see sections 5.2.6 and 5.2.7).

5.2.5   The Semantic Side of Resolving VP Ellipsis

Identifying the sponsor clause is just the start of resolving VP ellipsis. The process also involves answering up to five semantically oriented questions.

Semantic question 1   Do the subjects of the ellipsis clause and the sponsor clause corefer (5.36a) or not (5.36b)?

(5.36) a. Charlotte$_1$ wanted to **go paddleboarding** but she$_1$ couldn't __.
         b. Charlotte **wanted to go paddleboarding** but Laura didn't __.

In semantic terms, are the fillers of the external case-roles of the coreferential events coreferential? This decision informs the next decision.

Semantic question 2   Is there type- or instance-coreference between the events in the ellipsis clause and the sponsor clause? If the elided event refers to the same real-world instance as its sponsor, then there is instance-coreference (5.37a), whereas if the elided event refers to a different *instance* of the same *type* of event, there is *type-coreference* (5.37b).

(5.37) a. Jim tried to **open the bottle** but he couldn't __.
         b. Jim couldn't **open the bottle** but Jerry could __.

The main heuristic for identifying type- versus instance-coreference is that instance-coreference requires that all property values of the event unify.[18] In (5.37b), the OPEN events have different AGENTS, so they must be different instances.

Semantic question 3 If the sponsor clause contains modal and/or other matrix verbs (i.e., verbs that take verbal complements), are they included in, or excluded from, the ellipsis resolution? As the following pairs of examples show, the answer depends on the specific modal and matrix verbs used in the clauses as well as whether or not the subjects of the clauses are coreferential—all of which must be handled by a rule set that is consulted to answer semantic question #3.

(5.38) a. Whereas Nicholas promised to **help**, Alice actually did __.
    b. Whereas Nicholas **promised to help**, Alice didn't __.
    c. Whereas Nicholas promised to **help**, Alice didn't want to __.

(5.39) a. Nicholas said he would **come**, but he didn't __.
    b. Nicholas **said he would come**, but Alice didn't __.
    c. Nicholas said he would **come**, but Alice wouldn't promise to __.

(5.40) a. Nicholas **had to force himself to learn to skydive**, but Alice didn't __.
    b. Nicholas had to force himself to **learn to skydive**, but Alice really wanted to __.
    c. Nicholas **forced himself to learn to skydive**, but Alice couldn't __.

Semantic question 4 Is there type- or instance-coreference between VP-internal arguments and adjuncts in the ellipsis clause and the sponsor clause? Instance-coreference involves referring to the same actual entity multiple times (5.41), whereas type-coreference involves referring to different entities of the same type (5.42).

(5.41) I **jumped the fence into the alley**, and Sally did __ too. ; same fence and alley
(5.42) I **walk my dog every morning**, and Sally does __ too. ; different dog

Determining instance- versus type-coreference of objects can be tricky since it requires real-world reasoning. One clue comes from the determiners in the sponsor clause's noun phrases. Definite and demonstrative determiners tend to suggest instance-coreference (5.43), whereas indefinite determiners and possessive pronouns tend to suggest type-coreference (5.44 and 5.45).

(5.43) Julie **went to the basketball game**, and I did __ too. ; same game
(5.44) Julie **ate a sandwich**, and I did __ too. ; different sandwich
(5.45) Julie **washed her car**, and I did __ too. ; different car

Semantic question 5 Should adjuncts in the sponsor clause be included in, or excluded from, the resolution? If the ellipsis clause specifies a value for some property, and if the sponsor clause also has a value for that property, then the sponsor clause's value is not copied during ellipsis resolution. For example, in (5.46), the interpretation of *today* fills the TIME slot in the TMR for the ellipsis clause, which blocks copying of the interpretation of *yesterday* from the sponsor clause.

(5.46) William **arrived by train** yesterday, and Daniel did __ today.

All other property values, such as the meaning of *by train* in (5.46), are copied over during ellipsis resolution, and, if applicable, the instance- versus type-coreference rules described previously are invoked. For example, in (5.47), the meaning of *by the pool* fills the LOCATION slot and uses the same instance of SWIMMING-POOL because *the pool* is a definite description.

(5.47) Grandma was **drinking wine by the pool**, and Grandpa was __ too.

Returning to figure 5.2, even if making these semantic decisions results in a fully resolved instance of VP ellipsis, the associated TMR is still passed on to the final two stages of processing because, on the one hand, it could contain residual analysis needs unrelated to the ellipsis and, on the other hand, the referring expressions still need to be grounded to agent memory. Examples of VP ellipsis that cannot be resolved by the end of stage 4 of language understanding are reconsidered during stages 5 and 6—Extended Semantics and Situational Reasoning—when more context and more knowledge are brought to bear.

5.2.6 VP Ellipsis in Extended Semantics

During Extended Semantics, the agent is no longer using syntax to inform semantics; it is reasoning based entirely on TMRs.[19] One kind of VP ellipsis that is handled at this stage occurs in standard dialog pairs like the following.

(5.48) ["Do you **VP**? I don't __."]
"Do you **understand these instructions**? I don't__."

(5.49) ["Do you **Modals VP**?" "Yes, I do __. / No, I don't __."]
"Do you **want to collaborate**?" "Yes, I do __."

(5.50) ["Pronominal-Subj **VP**." "Who did __?"]
"They **punished him**." "Who did __?"

(5.51) ["**VP**$_{\text{Imperative}}$." "I don't want to __."]
       "**Get going**." "I don't want to __."

(5.52) ["I **see**." "No, you don't __."]
       "I **see**." "No, you don't __."

Although these constructions are presented as text strings for clarity's sake, they are actually recorded and processed in terms of meaning representations, which allows for both lexical variability (paraphrasing) and the potential for intervening material, as in (5.53).

(5.53) "I think I **understand**," said Rebecca. Shaking her head, Natasha replied, "No, you don't__."

Another kind of VP ellipsis that is handled during Extended Semantics is what we call *repetition structures*, which convey emphasis in utterances like the following:

(5.54) I can't **live without a potato peeler**. I just can't __.
(5.55) You don't **want to hang around with those people**. You really don't __.
(5.56) He didn't **know**. He honestly didn't __.
(5.57) Iris must **know about the situation**; of course she must __.

Like dialog strategies, repetition structures are best analyzed in terms of meaning representations, not language strings. The TMRs for each component are generated in the regular way during Basic Semantic Analysis, and the coreferences between the subjects and verb phrases are established during Lexically Triggered Procedural Semantics. Then, during Extended Semantics, the agent recognizes the repeated TMR chunks and merges them into a single one, decorated with the feature "EMPHASIS yes." If the second clause includes an adverb that, itself, conveys the meaning "EMPHASIS yes"— like *really* in (5.55)—then that feature value is listed only once in the merged TMR. If the second clause includes a modifier with a different meaning—like *honestly* in (5.56) and *of course* in (5.57)—then the analysis of the modifier is added to the merged TMR. In short, the *meaning* of the repetition is expressed by the merged TMR, the fact that this meaning was conveyed using repetition is not (though metadata attached to the TMR includes both the original input and the fact that the repetition-analysis rule fired).

5.2.7   VP Ellipsis in Situational Reasoning

Finally, during Situational Reasoning, the agent carries out many processes related to coreference and reference using its full toolbox of domain knowledge and situational awareness, which can include vision processing, mindreading of its interlocutor, reasoning about its own goals and plans, and so on. These processes include vetting candidate sponsors posited earlier, identifying sponsors for referring expressions that are as-yet unresolved, and anchoring all referring expressions—along with the information describing them—in the agent's memory.

5.3   Other Referring Expressions

The stagewise approach to treating coreference shown in the generic algorithm in figure 5.1 applies to all coreference phenomena. For reasons of space, we will illustrate its theme-and-variations nature using just two more of the many other kinds of referring expressions: overt event anaphors and personal pronouns.[20]

5.3.1   Event Anaphors

VP ellipsis is one of several ways to express event coreference.[21] Another is using overt anaphors like *do it*, *do this*, *do that*, and *do so*. Although, in some contexts, ellipsis and an overt anaphor can be used interchangeably, more often there is a clear preference, which points to different licensing conditions and norms of usage. For example, whereas VP ellipsis usually has a textual sponsor, event anaphors often refer to meanings that must be reconstructed using real-world reasoning, as shown by the following examples.

(5.58)   [*Do it* means *go to a fancy restaurant*]
"She'll still only eat at McDonald's and Sizzler," Shapiro adds. "If I even try to take her to a fancy restaurant, she won't **do it**."(COCA)

(5.59)   [*Do it* means *work on solving the problem of flooding in New Orleans*]
The government is getting ready to dump $200 billion into this problem, but not a . . . dime should be put into re-creating what was done in the past. Usually in American disasters such as this, we see knee-jerk reaction and brute force. But if we **do it** smart, we can in fact provide the protection that is needed.(COCA)

(5.60)   [*Do it* means *rebuild the Mississippi coast*, which is the topic of the article]
"The architectural heritage of Mississippi is fabulous, . . . really, really marvelous," Duany says, referring to antebellum mansions in Greek Revival and Federal styles that have imposing entrances, balconies and columns and smaller Creole cottages for the less wealthy. "However, what they have been building the last 30 years is the standard, tawdry strip developments. The government's vision is to start again and **do it** right."(COCA)

Since we happened to have developed the microtheory of VP ellipsis before working on overt event anaphors, the question was whether we could reuse that microtheory for event anaphors. The answer is "yes" but with narrower coverage because, in general, ellipsis is a more constrained process than the use of overt anaphors. This makes sense: if you're going to leave something out, you'd better be sure that the listener can both detect its absence and reconstruct its meaning. One requirement of VP ellipsis is that it be licensed by a modal word, which is not required of overt event anaphors. So, if we directly apply the modal-inclusive ellipsis rules to event anaphors, they will necessarily cover a smaller percentage of examples since they will ignore all examples in which the anaphor's clause lacks a modal word. For example, the ellipsis rules will cover *He felt like **cutting class** but he <u>couldn't</u> **do it***, but they will not cover *He felt like **cutting class** and so he **did it***. By contrast, if the ellipsis rules are modified so that they do not require a modal word in the anaphor's clause, they cover more examples but are less reliable, as became clear through informal experimentation. This suggests that modality is playing a larger role in establishing event coreference than might be obvious at first blush.

Below are examples of constructions in which verbal anaphors work the same as elided VPs. The constructions are categorized by the function words whose meaning-procedures zones hold the required anaphor-resolution function: conditional constructions are anchored in senses of *if*; clausal coordinate constructions are anchored in senses of *and* and *but*; and clausal juxtaposition constructions are anchored in senses of punctuation marks. In some cases, a construction that could formally be subsumed by another one is listed separately because of its expected frequency and the utility of storing, rather than dynamically computing, its analysis. For ease of reading, variations on the constructions are not presented.

Conditional Constructions

(5.61) [If Pro$_1$ PairedModals **ANA$_2$**, Pro$_1$ PairedModals **ANA$_2$**]
If we could **do it** with water, we would **do it** with water.(COCA)

(5.62) [If anybody can **ANA$_1$**, Pro can **ANA$_1$**]
But if anybody can **do it**, you can **do it**.(COCA)

(5.63) [If Pro$_1$ PairedModals **VP$_2$**, Pro$_1$ PairedModals **ANA$_2$**]

   a. But if they want to **buy services separately**, they can **do that**, too.(COCA)

   b. If we could **persuade our son**, we would gladly **do it**.(COCA)

(5.64) [If Subj$_1$ PairedModals **VP$_2$**, Pro$_1$ PairedModals **ANA$_2$**]
"If the political leaders could have **pulled those jobs off themselves**, they would have **done it**," he said.$_{(COCA)}$

(5.65) [If Subj$_1$ Modals **VP$_2$**, Pro$_1$ could at least **ANA$_2$**]
If Dole must **admit to linguistic chauvinism**, he could at least **do so** for the right reasons.$_{(COCA)}$

(5.66) [If Subj$_1$ was/were going to **VP$_2$**, Pro$_1$ would have **ANA$_2$**]
If they were going to **snap**, they would have **done it** before that.$_{(COCA)}$

Clausal Coordination

(5.67) [Pro$_1$ Modals$_3$ **VP$_2$** Punct and/but Pro$_1$ Modals$_3$ **ANA$_2$**]

   a. We're going to **break you**, and we're going to **do it** in three days.$_{(COCA)}$

   b. Somehow, and quick, he had to **maneuver his way to the opposite end, behind the receiving team, the Chicago Bears**. And he had to **do it** in such a way as to salvage the dignity of the always-suspect men in the striped shirts.$_{(COCA)}$

   c. He will **seek help on reconstruction in Iraq and reform in the Middle East**. And he will **do it** on a continent where he personally and his policies remain unpopular.$_{(COCA)}$

   d. We're going to **solve this case**, but we're going to **do it** our way.$_{(COCA)}$

   e. We want to **do something about it** but want to **do it** responsibly.$_{(COCA)}$

(5.68) [Subj$_1$ Modals$_3$ **VP$_2$** Punct and/but Pro$_1$ Modals$_3$ **ANA$_2$**]
Supervisors will **confront only the most overt offenders**, and they'll **do it** individually and in person.$_{(COCA)}$

(5.69) [Subj$_1$ Modals$_3$ **VP$_2$** Punct and Subj$_2$ Modals$_3$ **ANA$_2$**]
"A lot of Hollywood stars can **turn it on and off**, and athletes can **do that**, too," says Trampler.$_{(COCA)}$

(5.70) [Subj$_1$ wants to **VP$_2$** Punct but Pro$_1$ can't **do it$_2$** by Reflexive-pro]
We want to **keep the plant open**—but we can't **do it** by ourselves.$_{(COCA)}$

Clausal Juxtaposition

(5.71) [Subj$_1$ Modals$_3$ **ANA$_2$** Punct Pro$_1$ Modals$_3$ **ANA$_2$**]

   a. We don't want to **do that**. We've never wanted to **do that**.$_{(COCA)}$

   b. "And why did they have to **do this**? They had to **do it** to survive."$_{(COCA)}$

   c. Most of the time they will **do that**. They will **do it** in good faith.$_{(COCA)}$

   d. "We don't want to **do that**. We've never wanted to **do that**."$_{(COCA)}$

(5.72) [Subj$_1$ Modals$_4$ **ANA$_3$** Punct Subj$_2$ Modals$_4$ **ANA$_3$**]
        You would **do it**. Anybody would **do it**, and hopefully judges have enough
        integrity to administer the law fair-handedly.$_{(COCA)}$

In summary, the kinds of constructions that are useful for resolving VP
ellipsis are also useful for resolving verbal anaphors. However, the ellipsis-
oriented constructions cover a smaller percentage of inputs with verbal
anaphors because anaphors do not need to be licensed by a modal word.
The reason we do not relax the modal-word requirement for anaphoric
examples is that testing has shown that doing so significantly reduces the
sponsor-predicting power of the constructions.

5.3.2 Personal Pronouns

The model for treating personal pronouns, which closely parallels the models of VP ellipsis and verbal anaphors, is shown in figure 5.7.

Unlike the models of VP ellipsis and event anaphors, this one includes
a data-driven coreference engine, which suggests sponsors for many of
the personal pronouns encountered in texts. Its coreference links are
evaluated, along with those that the LEIA generates, during Situational
Reasoning.

The lexical senses for personal pronouns record a basic semantic analysis
in the sem-struc and include a call to a meaning procedure that attempts to
resolve the coreference.

he-pro1
  definition   the pronoun referring to a human male[22]
  example   He works at the library.
  syn-struc
   $var0
  sem-struc
  HUMAN-1
    HAS-GENDER male
  meaning-procedures
  resolve-coref-he HUMAN-1

The meaning procedures for different personal pronouns are largely the
same but have some noteworthy differences: for example, *he* can be generic,
she can refer to some inanimates, like boats, and *they* can require dynamic
construction of a set from multiple constituents. Pronouns can also participate in idiomatic expressions: for example, in *He who hesitates is lost*, *he* is

generic. If an input matches a fixed construction, shown in the left-hand path of figure 5.7, that analysis outscores compositional analyses that use the bare-pronoun sense. Everything else about processing personal pronouns follows the strategies presented earlier.

5.4 Porting the VP Ellipsis Model to Russian

As the last century of linguistic study has shown, languages are more alike than different. So, computational cognitive models of language should be informed by, and applicable to, different languages given appropriate parameterization. Having multiple languages in one's toolbox offers a modeler three advantages:

1. It can lead to scientific generalizations.
2. It invites evidence from corpora in multiple languages.
3. It fosters modeling that is maximally language-agnostic, such that models can be largely reused across languages given certain parameterization. This significantly reduces costs for ramping up systems in multiple languages.

For readers who might choose to skim this section, the main points are these:

1. Ellipsis is difficult to detect in corpora since one is looking for what is not there. This makes it hard to compile a large, truly representative research corpus. Since some kinds of elliptical examples are easier to find in English and some kinds are easier to find in Russian, there is a benefit to using both languages to create a research corpus.
2. In both Russian and English, overt anaphors can be used in some of the same contexts as ellipsis, which further expands the search space for useful examples.
3. Our approach to storing and using knowledge to resolve verb phrase ellipsis applies identically to English and Russian.

The microtheory of VP ellipsis in English can be largely reused for Russian with parameterization that involves case marking, word order, impersonal constructions, the inventory of auxiliaries, the expression of conditionals, and the ellipsis of other categories.[23]

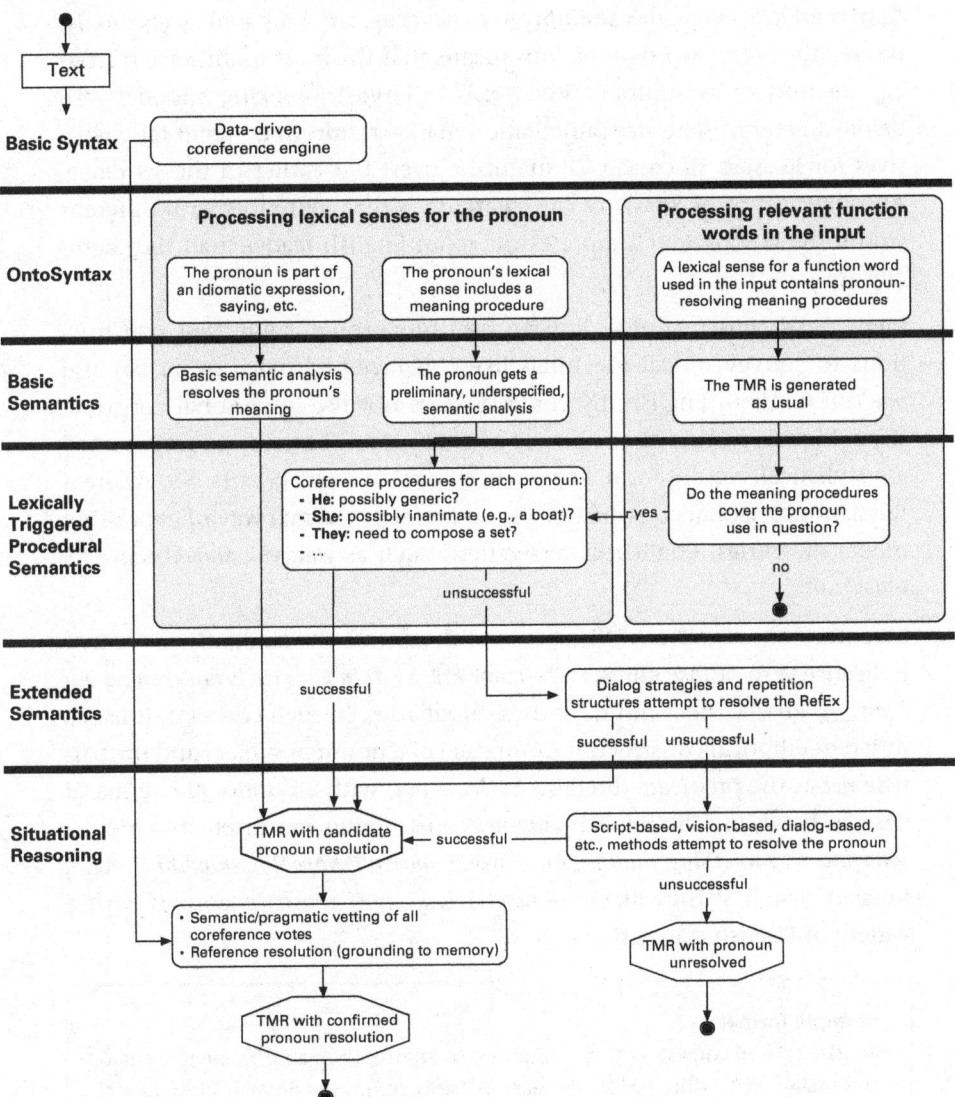

Figure 5.7
Top level of the algorithm for resolving personal pronouns.

Case marking Russian uses morphological case marking and, accordingly, has relatively free word order. This means that the most useful search strategy for finding examples of VP ellipsis in English—seeking a modal verb before an end-of-sentence punctuation mark—results in frequent false positives for Russian since the VP might be overt but earlier in the sentence. Another upshot of Russian's case marking is that sentences with different words orders look much more exotic to an English reader than they actually are.

Impersonal constructions Russian uses impersonal, subjectless constructions to convey modal meanings like *it is necessary* (*nado* or *nužno*) and *one can* (*možno*). English, by contrast, does not use impersonal constructions. Moreover, English *it-* and *one*-constructions like those just mentioned are stylistically more formal than their Russian counterparts. So, whereas Russian impersonal constructions are a perfectly normal way of expressing modal meanings, English prefers generic such as *you*: *you should <have to, must, can>.*

Inventory of auxiliaries Whereas English can use the auxiliaries *do* and *be* to license verb phrase ellipsis (*She **came** but we didn't __; He is not **coming** but I am __*), Russian does not have these auxiliaries. In such contexts, it uses a different elliptical construction composed of a nominal subject and the particle *net* as the predicate (literally, Subj$_{NOM}$ *not*) with an optional comma or dash in between. The Russian *Subj-net* construction can imply any modal: *someone or something didn't <can't, won't, doesn't have to, shouldn't,* etc.>. Russian search strings like *On—net* (He$_{NOM}$—*not*) return examples with a variety of English translations.

Example format
For the sake of consistency, all examples, no matter their source, are presented as Russian in Cyrillic script, Russian in Latin script, an English gloss of the Russian, and finally an English version. A subscript indicates the source of the example. Only select grammatical information is presented.

(5.73) «Вы **что-нибудь понимаете**? Я—нет». (Russian National Corpus)
«Vy **čto-nibud' ponimaete**? Ja—net».
You$_{NOM}$ **something**$_{ACC}$ **understand**? I$_{NOM}$—not
"Do you **understand what's going on**? I don't__."

(5.74) Я смог **выдержать до конца**, а он нет. (SketchEngine: Russian)
Ja smog **vyderžat' do konca**, a on net.
I_{NOM} could **hold-out to end**, but he$_{NOM}$ not
I could **hold out till the end**, but he couldn't __.

(5.75) Летчик должен был **отказаться от взлета**, но он этого не сделал. (SketchEngine: Russian)
Letčit dolžen byl **otkazat'sja ot vzleta**, no on ètogo ne sdelal.
Pilot$_{NOM}$ should have **refused Prep taking-off**, but he$_{NOM}$ that$_{GEN}$ not did
The pilot should have **refused to take off**, but he didn't __.

Expression of conditionals Russian forms conditionals using the particle *by*, which cannot license verb phrase ellipsis. So, for English sentences in which *would* licenses VP ellipsis, Russian must use a different strategy, such as repeating the verb or using an overt anaphor. Surface mismatches like these are useful for language modeling because they point out related phenomena that one might not think of otherwise. For example, in (5.76) Russian repeats the verb, a strategy that is possible in English as well: *If I could have caught him, I would have caught him.*

(5.76) Если бы я мог **его поймать**, я бы поймал __.
Esli by ja mog **ego pojmat'**, ja by pojmal __.
If Condit I_{NOM} could **him$_{ACC}$ catch**, I_{NOM} Condit caught __
I mean, if I could have **caught him**, I would have __. (COCA)

Ellipsis of other categories Russian permits subjects and objects to be elided under certain conditions. For example, in (5.77) and (5.78), the clause with VP ellipsis also has subject ellipsis.

(5.77) **Убей меня сейчас**, если можешь __. (Искандер)
Ubej menja sejčas, esli možeš' __. (Iskander)
Kill$_{IMPER}$ me$_{ACC}$ now if can$_{2.SG}$ __
Kill me now if you can __.

(5.78) «Ты **хочешь продолжать**?»—«Хочу __».
«Ty **xočeš' prodolžat'**?»—«Xoču __».
You$_{NOM}$ **want-to continue**?—Want-to$_{1.SG.PRES}$ __
"Do you **want to proceed**?" "Yes, I do __." (COCA)

We are not suggesting that someone barely familiar with a second language would find this bilingual methodology useful or that the lockstep treatment of every found example for both languages is appropriate. Instead, our experience has shown that multilingual analysis is useful as long as the knowledge engineer is well-versed in all languages consulted.

Tools   Different corpora and different search tools are available for different languages. In this research, we used the online search capabilities of The Corpus of Contemporary American English (COCA; Davies, 2008–), the Russian National Corpus (RNC; ruscorpora.ru), and Sketch Engine (Kilgarriff et al., 2014), as well as offline versions of COCA and the Gigaword Corpus (Graff & Cieri, 2003). While online search engines offer useful functionalities, they do not offer everything that one can include in a customized search program. For example, using the abovementioned engines, it is not possible to batch search for any modal verb within some number of words of its negative counterpart. However, this can actually be worked around since different modal pairs tend to work largely the same when it comes to modeling ellipsis resolution, so orienting around a few specific pairs is likely to reveal the majority of kinds of contexts that need to be treated in the model.

In sum, Russian and English are typologically similar with respect to the parameters that serve as heuristic evidence for the VP ellipsis model described here, and those differences are readily accommodated by our modeling strategy. This means that the implemented model for English is largely reusable for Russian.

6   Dialog as Perception, Deliberation, and Action

Since LEIAs are primarily intended to serve as members of human-agent teams, they must be adept at communicating in the way that is most natural for people—using language. This chapter describes how LEIAs participate in dialog, which illustrates their flow of Perception, Deliberation, and Action. No separate dialog model or dialog manager is needed. Instead, dialog is treated within a model of multimodal interaction in which utterances have the same status as nonspeech events. A key modeling strategy is the use of scriptlets, which are pairs of ontological concepts that record typical sequences of events. For example, when someone asks a question, the default next move is for the interlocutor to answer it. In developing this microtheory, priorities include, as always, simpler-first modeling, strategic knowledge engineering, maximum domain-independence, and overall transparency, thus enabling explainability and system enhancement over time.

6.1   The Tradition of Dialog Modeling

Human interactions follow standard patterns: questions are typically followed by answers, requests are typically followed by an agreement or refusal to carry them out, and so on. Such pairs of moves are called *adjacency pairs*. Of course, other things can happen as well, such as responding to a question by asking for clarification or by stomping off in a huff. Both the default actions and the less common ones are part of people's knowledge about the world, so they must be part of an agent's knowledge as well.

There is a long history of scientific work on dialog, but it only lightly informs LEIA modeling. To show why, we will briefly describe the objectives of, and research results from, four thrusts of work on dialog.[1]

Dialog in pragmatics Pragmatics is the branch of linguistics that studies language in context. Within pragmatics—specifically, noncomputational pragmatics—linguists have produced taxonomies of speech acts and descriptions of how dialog works. However, these taxonomies and descriptions are limited in three ways.

1. They only address the *functions* of utterances in dialogs, such as making a request or asking a question, not the *meanings* of the utterances. For example, *Please make me a latte* and *Please change the oil in my car* mean completely different things even though they are both requests for action.

2. The taxonomies are not implementable because they do not provide heuristics that would allow a computer system to automatically detect the different kinds of speech acts. In fact, they cannot do that even in principle because a speech act cannot be detected without knowing what the utterance means. This is because the form of an utterance does not always align with its function. For example, a declarative sentence can make an assertion (*It's hot out*), request information (*I need to know your mailing address*), request action (*I could use some help here*), or make a threat (*I wouldn't do that if I were you*).

3. The taxonomies isolate language from all other communicative and extra-communicative acts despite the fact that, in real life, verbal and nonverbal actions freely interact. For example:
   - When X asks Y a question, Y can speak an answer, shrug, or point to the answer in the environment.
   - When X tells Y a joke, Y can laugh, smile and say it's funny, or look puzzled (he didn't get it).
   - When X punches Y, Y can punch him back, yell at him, or run.
   - When X asks Y to do something, Y can say she'll do it (but not yet do it), say she'll do it and also start doing it, or say nothing and just set off to do it.

As with descriptive linguistics overall, the main use of pragmatic accounts of dialog for cognitive modeling is to jog knowledge engineers' memories about the phenomena that need to be covered.

Dialog in early knowledge-based natural language processing Before the statistical turn of the 1990s, computational linguists worked on dialog models that are in the spirit of what we advocate.[2] However, the models largely isolated dialog from nonlinguistic actions while acknowledging the need for future multimodal expansion. We, by contrast, consider it strategically imperative to consider the big picture first in order to avoid building models, and associated system modules, that will need to be redesigned when the agent faces the real-world state of affairs, such as multimodal interaction. Early knowledge-based work on dialog stopped around the turn of the century, presumably due to the scientific and extra-scientific pressures of the statistical turn.

Dialog in narrow-domain AI Within narrow-domain AI systems, dialog acts are commonly paired with the specific propositions needed in the application. For example, Jeong and Lee's (2006) flight reservation system considers "Show Flight" to be a dialog act even though it is actually a combination of a request for action, which is a proper dialog act, and the meaning of that action—showing a flight. Although creating idiosyncratic pairs of dialog acts and their meanings is an efficient way to ramp up narrow-domain systems, this kind of work is not extensible to other domains and does not offer a path toward agent systems that can operate across domains.

Dialog in data-driven AI Developers of data-driven AI systems have used dialog-act tags as features to inform supervised machine learning.[3] Although the inventory of tags can be theoretically informed, the tag set that is ultimately used for a particular application is whatever works best, based on system testing. For example, Stolcke et al. (2000) used forty-two dialog acts to improve speech recognition in the Switchboard corpus of human-human conversational telephone speech.[4] By contrast, the Map Task used twelve speech acts, none of which overlap with the Verbmobil-1's dozens of speech acts (Jurafsky, 2006). Using bespoke feature sets improves system performance and gives applications the veneer of linguistic grounding, but such approaches do not contribute to either language understanding or generalized agent capabilities.

In conclusion, taxonomies and lists can be useful fodder for knowledge engineers, in a similar way as are human-oriented dictionaries, grammars, and thesauri. However, to design a knowledge-based, domain-independent, multimodal model of communication for LEIAs, we essentially had to start from scratch.

6.2   Communicative Acts: Events Like Any Others

Embodied human communication combines spoken language, body language, and general actions that, in a given context, serve to communicate. For example, you can request that another person come along by speaking (*Come with me*), waving your arm in the direction you're walking, or just setting off if you've previously agreed to stick together. Accordingly, a LEIA's ontological knowledge about communication prioritizes *what* is communicated over *how* it is communicated. The *how* is taken care of at the flanks of the cognitive architecture: on the input side, by the agent's perception processing engines, and on the output side, by the agent's decision-making about how to carry out whatever action it decides to take.

Below is a short excerpt from the COMMUNICATIVE-ACT subtree of the ontology that focuses on asking questions (REQUEST-INFO) and answering them (RESPOND-TO-REQUEST-INFO).

REQUEST-INFO
   REQUEST-INFO-YN
   REQUEST-INFO-WH
     REQUEST-PROBLEM-DESCRIPTION
      REQUEST-MEDICAL-COMPLAINT
      REQUEST-MECHANICAL-PROBLEM
      REQUEST-ISSUE-OF-CONSULTATION
RESPOND-TO-REQUEST-INFO
   RESPOND-TO-REQUEST-INFO-YN
     RESPOND-TO-REQUEST-INFO-POSITIVELY
     RESPOND-TO-REQUEST-INFO-NEGATIVELY
     RESPOND-TO-REQUEST-INFO-DON'T-KNOW
   RESPOND-TO-REQUEST-INFO-WH
     DESCRIBE-PROBLEM
      DESCRIBE-MEDICAL-COMPLAINT
      DESCRIBE-MECHANICAL-PROBLEM
      DESCRIBE-ISSUE-OF-CONSULTATION

We will use this ontology excerpt to illustrate some key points about how communication is modeled. Asking questions (REQUEST-INFO) is divided into subclasses for the following reasons:

1. This mirrors human knowledge: people know that there are different kinds of questions that ask different kinds of things.

2. Responding to different kinds of questions requires different kinds of reasoning that results in different kinds of responses. For example, when a doctor asks a virtual patient about its medical complaint (REQUEST-MEDICAL-COMPLAINT), the agent needs to search its memory for symptoms and decide which ones to report and in how much detail. The reasoning function for that is attached to the concept DESCRIBE-MEDICAL-COMPLAINT.

3. The question-answer pairs are linked using the property ADJACENCY-PAIR. For example, the adjacency pair of REQUEST-MEDICAL-COMPLAINT is DESCRIBE-MEDICAL-COMPLAINT. Adjacency pairs guide the LEIA's next move when it is participating in a dialog, and they also help it to interpret others' moves when it is observing a dialog. Recall (from section 3.2.4) that the agent knows how to leverage adjacency-pair information thanks to the AGENT-FUNCTIONING-FLOW script, which implements its cognitive architecture.

4. Different language expressions and gestures convey specific kinds of questions and answers, and this information is recorded in the LEIA's lexicon and opticon.

The following is a larger excerpt from the COMMUNICATIVE-ACT subtree (not all subtrees are fully expanded).

COMMUNICATIVE-ACT
  REQUEST-INFO
    REQUEST-INFO-YN
    REQUEST-INFO-WH
      REQUEST-PROBLEM-DESCRIPTION
        REQUEST-MEDICAL-COMPLAINT
        REQUEST-MECHANICAL-PROBLEM
        REQUEST-ISSUE-OF-CONSULTATION
  RESPOND-TO-REQUEST-INFO
    RESPOND-TO-REQUEST-INFO-YN
      RESPOND-TO-REQUEST-INFO-POSITIVELY
      RESPOND-TO-REQUEST-INFO-NEGATIVELY
      RESPOND-TO-REQUEST-INFO-DON'T-KNOW
    RESPOND-TO-REQUEST-INFO-WH
      DESCRIBE-PROBLEM
        DESCRIBE-MEDICAL-COMPLAINT
        DESCRIBE-MECHANICAL-PROBLEM
        DESCRIBE-ISSUE-OF-CONSULTATION
  REQUEST-ACTION
  RESPOND-TO-REQUEST-ACTION

PROPOSE-PLAN
  PROPOSE-MEDICAL-INTERVENTION
RESPOND-TO-PROPOSED-PLAN
  RESPOND-TO-PROPOSED-MEDICAL-INTERVENTION
SPEECH-ACT
  DECLARATIVE-SPEECH-ACT
  TELL-A-JOKE
  EMPTY-CONTENT-SPEECH-ACT
PERFORMATIVE-ACT
  APOLOGIZE
  EXPRESS-DOUBT
  PRAISE
  THREATEN
EMOTIONAL-COMMUNICATIVE-ACT
  EXPRESS-DISPLEASURE
  EXPRESS-PLEASURE
  EXPRESS-UNCERTAINTY
BACKCHANNEL-SIMPLE
CHECK-UNDERSTANDING
SEEK-CLARIFICATION
HEDGE-BUY-TIME

Some noteworthy features of the COMMUNICATIVE-ACT subtree are as follows:

- The only subtree in which the communication must involve language is the SPEECH-ACT subtree.

- The way that a COMMUNICATIVE-ACT is typically conveyed is recorded using the property INSTRUMENT. For example, the description of PRAISE specifies that the INSTRUMENT is typically SPEECH-ACT, APPLAUD, or GIVE-THUMBS-UP.

- Many of these concepts have a default adjacency pair, though it is not always a COMMUNICATIVE-ACT. For example, the default response to someone telling a joke (TELL-A-JOKE) is laughing (LAUGH), which is an EMOTIONAL-EVENT.

- This subtree, like all aspects of ontology, will evolve over time in response to actual agent needs. For example, the concepts REQUEST-MEDICAL-COMPLAINT and DESCRIBE-MEDICAL-COMPLAINT exist because the virtual patients in a medical application needed to understand related questions and make decisions in response to them.

We now turn to how agents use their ontological knowledge about communication when participating in dialog.

6.3   Examples of Dialog as Perception, Deliberation, and Action

We will illustrate dialog processing using LEIAs that are playing the role of virtual patients in a clinician training environment. The application mirrors the previously reported Maryland Virtual Patient system (McShane & Nirenburg, 2021, chapter 8), but the modeling reflects recent advances in generalizing all aspects of agent functioning beyond the needs of individual demonstration systems.

6.3.1   The Doctor Asks, "What brings you here?"

When the doctor asks, "What brings you here?", the LEIA virtual patient creates a new instance of the AGENT-FUNCTIONING-FLOW script (cf. section 3.2.4). As a reminder, this script has five subevents:

AGENT-FUNCTIONING-FLOW
   SUBEVENTS
     1. PERCEPTION-RECOGNITION
     2. PERCEPTION-INTERPRETATION
     3. DELIBERATION
     4. ACTION-SPECIFICATION
     5. ACTION-RENDERING

The agent recognizes *What brings you here?* as a language input and then uses its language understanding engine to interpret its meaning. That involves the six-stage process described in chapter 4, highlights of which we work through below.

The construction *What brings you here?* is polysemous. In its broadest sense, it expresses mild surprise at seeing somebody somewhere and asks why they came. However, in different service contexts, it has more specific connotations. If a mechanic says this to a car owner, it means *What's wrong with your car?* If a professor says this to a student showing up for office hours, it means *What do you need my help with?* And, if a doctor says this to a patient, it means *What is your medical complaint?* The best way to model these variations is by recording the generic meaning of the construction in the sem-struc of the lexical sense for *What brings you here?* (what-n16) and appending meaning procedures that can dynamically detect whether a narrower interpretation is warranted by the context.

what-n16
    definition   The construction "What brings you here?," which is an informal paraphrase of "Why did you come here?"
    example   What brings you here?
    comments   Specialist interpretations are handled by conditions recorded as meaning procedures.
    syntax-type   atypical
    output-syntax   CL
    syn-struc
      subject $var0 (root 'what')
      v $var1 (root 'bring') (tense present)
      directobject $var2 (root 'you')
      adv $var3 (root 'here')
      punct $var4 (root question-mark) (opt+)
    sem-struc
      REQUEST-INFO
        AGENT   *speaker*
        BENEFICIARY   ^$var2
        THEME   COME-1.CAUSED-BY
      COME-1
        AGENT   ^$var2
        DESTINATION   ^$var3
      ^$var1,4 null-sem+

    meaning-procedures
      TRY: detect-doctors-appointment   ; changes analysis to REQUEST-MEDICAL-COMPLAINT
      TRY: detect-mechanical-service-event   ; changes analysis to REQUEST-MECHANICAL-PROBLEM
      TRY: detect-conversational-consultation ; changes analysis to REQUEST-ISSUE-OF-CONSULTATION

Each of the conditions listed as meaning procedures detects a particular kind of speaker, addressee, and location. If the conditions for a particular meaning procedure are met, that function replaces the generic interpretation with a more specific one. For example, in order for the medical complaint interpretation to be used, the speaker must be a doctor, the addressee must be a patient, and the location must be a medical building. If these conditions hold, then the replacement interpretation is "REQUEST-MEDICAL-COMPLAINT (AGENT DOCTOR) (BENEFICIARY PATIENT)." By contrast, if a doctor says, "What brings you here?" to a patient she sees at a soccer game, the more generic interpretation is retained. Since meaning procedures are run during the natural language understanding per se (specifically, during stage 4, Lexically Triggered Procedural Semantics) the text meaning representation (TMR) the agent generates as the result of Perception Interpretation already reflects the above reasoning.

The next subtask in AGENT-FUNCTIONING-FLOW is DELIBERATION, which uses this TMR as input. Since REQUEST-MEDICAL-COMPLAINT has an adjacency pair, the LEIA's default action is to instantiate the paired concept: DESCRIBE-MEDICAL-COMPLAINT.

An important detail about the ontological description of adjacency pairs is that they indicate how case-role fillers correlate across the paired events. For our example, the patient who is asked about his or her medical complaint is the one who will describe it, a coreference that is shown using the coreference slot.[5]

DESCRIBE-MEDICAL-COMPLAINT

AGENT	sem	PATIENT
	coref	REQUEST-MEDICAL-COMPLAINT-1.BENEFICIARY
BENEFICIARY	sem	DOCTOR
	coref	REQUEST-MEDICAL-COMPLAINT-1.AGENT
THEME	sem	PATHOLOGIC-FUNCTION
ADJACENCY-PAIR-OF	default	REQUEST-MEDICAL-COMPLAINT
	coref	REQUEST-MEDICAL-COMPLAINT-1

SUBEVENTS
   1. SEARCH-MEMORY-SYMPTOMS
   2. SELECT-REPORTABLE-SYMPTOMS
   3. SELECT-REPORTABLE-SYMPTOM-DETAILS

The three subevents of this script are, themselves, scripts that require specific reasoning that can be modeled at varying levels of complexity.

1. SEARCH-MEMORY-SYMPTOMS. Since not all symptoms are experienced continually, the question is not necessarily about the agent's symptoms at the moment the question is asked but, rather, about the recent pattern of symptoms. Making the necessary generalization can be straightforward or complicated depending on how a disease is modeled as well as the agent's approach to memory management. If symptoms are modeled using generalizing attributes—like HEARTBURN-LEVEL and DIFFICULTY-SWALLOWING, which are measured on the abstract scale {0,1}—this directly provides the needed generalization. For example, if the patient searches its memory and finds "HEARTBURN-LEVEL .5," this means that it is experiencing moderate heartburn. (This is how symptoms were modeled in the Maryland Virtual Patient system.) If, by contrast, no generalizing attributes are included in the disease model, then the agent needs to generalize about past and current symptoms on the fly.

2. SELECT-REPORTABLE-SYMPTOMS. Patients can have symptoms that are relevant for different medical specialists. For example, when consulting a gastroenterologist, it would be inappropriate to report a recent toothache or a burnt finger. Furthermore, the agent might be experiencing more symptoms than a person would typically list at one go. It would sound unnatural to list seven symptoms in response to a doctor's opening question. It is much more natural, as an initial response, to select only the most important ones, which can be identified based on a combination of their intensity, duration, and potential risk. For example, chest pain is an alarm symptom that should always be reported.

3. SELECT-REPORTABLE-SYMPTOM-DETAILS. In order to behave in a humanlike way, the agent should avoid providing too many details about each symptom or reporting them in a strictly parallel fashion. The following would not merely sound robotic; it would be downright off-putting: *I have moderate heartburn that began on March 4th and occurs three times daily. And I have intense chest pain that began on April 20th and occurs every four hours. And I have . . .* A person would say something more like *I have sharp chest pain and heartburn too* and then let the doctor follow up with more questions.

The example of reporting one's medical complaint to a doctor underscores the following:

- Computationally modeling human-level decision-making is a formidable task, and work on cognitive systems will grind to a halt if we insist on human-level complexity and performance immediately and at every turn.
- Not every decision needs to involve maximally complex reasoning in order to create useful agent applications.
- More complexity does not necessarily lead to better decision-making or better models overall.[6]

Consider an example in which a simplified model for reporting symptoms would be entirely sufficient. In a clinical training system that features virtual patients, the patients might never experience more than four symptoms at once; they might never experience symptoms that are irrelevant to their major complaint; and they might all be modeled to be not very talkative in order to test trainees' ability to recall which follow-up questions need to be asked. Under such circumstances, the decision functions involving what to report and how can be simple: report all symptoms in recent memory and say nothing about them beyond their name. In saying that well-selected simplifications are useful for agent modeling, we do not want to imply that knowledge engineering should back off all complexity. Instead, we are only pointing out that good judgment is needed to identify where complexity will best contribute to applications and where it would be an unneeded flourish.

As a specific example of reporting symptoms, assume that, in a given simulation run, the agent proceeds through DESCRIBE-MEDICAL-COMPLAINT as follows:

1. SEARCH-MEMORY-SYMPTOMS results in identifying two symptoms, DYSPHAGIA and REGURGITATION.

2. SELECT-REPORTABLE-SYMPTOMS results in the decision to report both of them.

3. SELECT-REPORTABLE-SYMPTOM-DETAILS results in the decision to not report any details.

The MMR (mental meaning representation) resulting from this reasoning will be as follows:

SPEECH-ACT-1
 AGENT   LEIA-1
 THEME   SET-1
SET-1
 ELEMENTS   DYSPHAGIA-1, REGURGITATION-1
DYSPHAGIA-1
 EXPERIENCER LEIA-1
 TIME   *anchor-time*
REGURGITATION-1
 EXPERIENCER LEIA-1
 TIME   *anchor-time*

Generating this MMR is the last step in DELIBERATION. Next, the agent proceeds to ACTION-SPECIFICATION, where it converts the MMR into a GMR (generation meaning representation). Finally, it uses this GMR as input to ACTION-RENDERING, where it generates a sentence as a response, such as "I have difficulty swallowing and regurgitation."

The script for describing patient complaints does not cover lying or exaggerating, as this is not the place to handle those behaviors. Those need to be modeled as high-level behaviors that can manifest across domains.

This concludes the cycle of AGENT-FUNCTIONING-FLOW triggered by a doctor asking a LEIA virtual patient *What brings you here?* After the agent responds, it waits for the doctor to make the next move.

6.3.2 The Doctor Asks, "Do you have chest pain?"

Interpreting and responding to *Do you have chest pain?* starts out the same way as the last example. First the LEIA analyzes the input using its language understanding engine. This question includes two linguistic constructions that must be treated together: a *yes-no* question recorded as the lexical sense do-aux47, and the construction *someone has (N$_{BODY-PART}$) pain*, recorded as the sense have-v32.

do-aux47
    definition   the construction "Do/did Subj VP?"
    example   Do you swim every day? Did Nellie make this pie?
    syntax-type   atypical
    output-syntax   CL
    syn-struc
      aux $var0
      subject $var1 (subject of $var2)
      vp $var2
      punct $var3 (root quest-mark) (opt+)
    sem-struc
      REQUEST-INFO-YN
        AGENT   *speaker*
        BENEFICIARY   *hearer*
        THEME   MODALITY-1.VALUE
      MODALITY-1
        TYPE   EPISTEMIC
        SCOPE   ^var2
      ^$var3   null-sem+

Do-aux47 is read as follows:

- The syn-struc defines the construction syntactically and allows for any shape of subject and verb phrase, both of which can be arbitrarily complex.
- The sem-struc is headed by the communicative act REQUEST-INFO-YN, whose AGENT is the speaker and whose BENEFICIARY is the hearer; that is, the speaker asks the hearer this yes-no question.
- The THEME of REQUEST-INFO-YN is the heart of this construction: it is the as-yet unknown value of epistemic modality scoping over the proposition.

Epistemic modality captures the factivity of an event. For the current purposes, suffice it to say that a value of 1 means that an event is actual (the answer to the yes-no question is *yes*), and a value of 0 indicates negation (the answer to the yes-no question is *no*).

- The scope of the epistemic modality is the meaning of the proposition, which is headed by ^$var2.

- The meanings of $var1 and $var2 (the subject and predicate) are combined using generic processes, so ^$var1 need not be overtly specified in the sem-struc.

- The last line in the sem-struc zone indicates that the meaning of the question mark has already been taken care of—it need not be analyzed further.

To analyze *Do you have chest pain?*, the question-oriented construction above must be combined with the construction *someone has (N_{BODY-PART}) pain*, which is recorded as the lexical sense have-v32.

have-v32

definition	the construction "X has (body part) pain"
example	I have pain. She has knee pain.
syntax-type	atypical
comments	This construction folds in the analysis of the nominal compound and the fact that the body part belongs to the animal experiencing the pain.
output-syntax	CL
syn-struc	
subject	$var1
v	$var0
n	$var2 (opt+)
n	$var3 (root 'pain')
sem-struc	

  PAIN
    EXPERIENCER   ^$var1
    LOCATION   ^$var2 (sem BODY-PART)
  ^$var2
    PART-OF-OBJECT   ^$var1
  ^$var3   null-sem+

The sem-struc of this sense can be read as *someone experiences pain in a location that is his or her body part*. Being a basic verbal sense, have-v32 is declarative and gets transformed to accommodate the question during transformation processing in OntoSyntax (section 4.2.2). Using these and other lexical senses, the agent generates the following TMR for *Do you have chest pain?*

REQUEST-INFO-YN-1
    AGENT   HUMAN-1
    BENEFICIARY   LEIA-1
    THEME   MODALITY-1.VALUE
MODALITY-1
    TYPE   EPISTEMIC
    SCOPE   PAIN-1
PAIN-1
    EXPERIENCER   LEIA-1
    LOCATION   CHEST-1
    TIME   *find-anchor-time*
CHEST-1
    PART-OF-OBJECT   LEIA-1

As in the previous example, this TMR includes a concept, REQUEST-INFO-YN, that has an adjacency pair: RESPOND-TO-REQUEST-INFO-YN, which the agent instantiates as the plan for its next move. This plan largely parallels the one in the last example in that the agent needs to search its memory for the answer, decide what to report, decide how many details to provide, and generate a response. Assuming that the answer is positive—that is, the agent does have chest pain—the GMR will look as follows:

RESPOND-TO-REQUEST-INFO-POSITIVELY-1
    AGENT   LEIA-1
    CAUSED-BY   PAIN-1
PAIN-1
    EXPERIENCER   LEIA-1
    LOCATION   CHEST-1
    TIME   *anchor-time*
CHEST-1
    PART-OF-OBJECT   LEIA-1

This not only says that the response is positive but also indicates the reason for responding positively: the fact that the agent has chest pain. However, this representation does not indicate how the agent will ultimately choose to express this meaning. If the agent is embodied, either in a robot or in simulation, it could nod its head. The associated meaning-to-effector mapping would need to be recorded in an action-generation knowledge base analogous to the lexicon and opticon. If, by contrast, the agent chooses to speak the response, then it needs to look in its lexicon for senses that express RESPOND-TO-REQUEST-INFO-POSITIVELY and decide whether to provide a bare response, such as *Yes*, or a response that repeats the core content, such as *Yes, I do have chest pain*. By default, LEIAs are currently modeled to be terse. One of the lexical senses that conveys the meaning ANSWER-YN-QU-POSITIVELY is yes-adv2, which the language generator can use to generate a response.

yes-adv2

definition	Used as fragmentary response to a yes-no question.
example	"Did you have a cookie?" "Yes."
syntax-type	fragment
output-syntax	CL
syn-struc	
$var0	
sem-struc	

    ANSWER-YN-QU-POSITIVELY

      AGENT   *speaker*

      BENEFICIARY *hearer*

6.3.3 The Doctor Proposes a Medical Intervention

In the spirit of patient-centered medicine, doctors in the US are not supposed to tell patients what to do; they are supposed to make recommendations (propose plans) and negotiate with patients to find a mutually acceptable path forward. We will work through the example of a doctor recommending that the patient have the diagnostic procedure esophagogastroduodenoscopy (EGD) by saying "I'd like to set you up for an EGD". This example illustrates both reasoning about a proposed plan and opportunistic learning by the agent. Opportunistic learning is learning that occurs as a side-effect of pursuing another goal—in this case, getting diagnosed and treated.

The utterance "I'd like to set you up for an EGD" uses the construction *I'd like to set NP up for NP*, which is recorded in the lexicon as the sense like-v11. Its semantic interpretation uses the concept PROPOSE-MEDICAL-INTERVENTION.

like-v11

definition	The construction "I'd like to set NP up for NP."
example	I'd like to set you up for a Heller myotomy.
syntax-type	atypical
comments	Constrained to a doctor advising a patient to have a procedure. The constraints on the agent (doctor), beneficiary (patient or his/her guardian), and theme (medical procedure) do not need to be listed because they are asserted as case role constraints in the ontological concept PROPOSE-MEDICAL-INTERVENTION.

```
output-syntax    CL
syn-struc
   subject       $var1      (root 'I')
   aux           $var2      (root 'would')
   v             $var0
   xcomp         $var3      (root 'set')
   directobject  $var4
   prep-part     $var5      (root 'up')
   pp
     prep        $var6      (root 'for')
     obj         $var7
sem-struc
   PROPOSE-MEDICAL-INTERVENTION
     AGENT           ^$var1
     BENEFICIARY     ^$var4
     THEME           ^$var7
   ^$var2,3,5,6 null-sem+
```

Most agents will not know the term EGD, which will launch new-word learning as a matter of course during language understanding (cf. sections 4.2.2 and 7.1.1). Specifically:

- During Basic Syntax, the part-of-speech tagger recognizes EGD as a noun and the parser recognizes it as the object of the preposition *for*.

- During OntoSyntax, the LEIA generates a lexical sense for this unknown noun, mapping it tentatively to OBJECT.

- During Basic Semantics, one of the candidate interpretations of the sentence uses like-v11, which requires that the object of *for* ($var7) be a type of MEDICAL-INTERVENTION. In other words, if like-v11 reflects the correct way of analyzing *like* this context, then EGD must refer to a MEDICAL-INTERVENTION.

- Following this clue, the agent remaps EGD from OBJECT to MEDICAL-INTERVENTION in this candidate interpretation of the input.

- Then, when the agent engages in concept learning (cf. section 7.2.2), it learns the new concept EGD, remaps the word EGD from MEDICAL-INTERVENTION to EGD, and ultimately understands that the input *I'd like to set you up for an EGD* means:

```
PROPOSE-MEDICAL-INTERVENTION-1
   AGENT          DOCTOR-1
   BENEFICIARY    LEIA-1
   THEME          EGD-1
```

As in previous examples, this TMR includes a concept, PROPOSE-MEDICAL-INTERVENTION, that has an adjacency pair: RESPOND-TO-PROPOSED-MEDICAL-INTERVENTION. So, the agent instantiates this concept as the plan for its next move. That plan has four ordered subevents:

RESPOND-TO-PROPOSED-MEDICAL-INTERVENTION
  SUBEVENTS
    1. DETERMINE-IF-ENOUGH-INFO-TO-EVALUATE
    2. COLLECT-NEEDED-FEATURE-VALUES
    3. RUN-EVALUATION-FUNCTION
    4. FORMULATE-RESPONSE

The first subevent determines whether the agent has enough information about the procedure to make a decision. This depends on its character traits, which include, non-exhaustively:

TRUST-IN-DOCTOR	{0,1}
COURAGE	{0,1}
SUGGESTIBILITY	{0,1}
NEED-TO-KNOW-RISKS	yes/no
NEED-TO-KNOW-PAIN	yes/no
NEED-TO-KNOW-SIDE-EFFECTS	yes/no
NEED-TO-KNOW-DEFINITIONS	yes/no ; for unknown-word processing

Certain combinations of these feature values—such as high trust in the doctor, high courage, and high suggestibility—result in the patient immediately agreeing to anything the doctor suggests. By contrast, a positive value for any need-to-know property causes the agent to check if it knows the contextually relevant value; if not, it must undertake to learn it before making a decision.[7] These "needs to know" are modeled as daemons—essentially, standing plans that are triggered any time their preconditions are met (cf. section 8.4).

Let us consider the example of a virtual patient with average values of TRUST-IN-DOCTOR, COURAGE, and SUGGESTIBILITY (all .5), a need to know the definitions of unknown words (NEED-TO-KNOW-DEFINITIONS yes), and a need to know the risks of interventions (NEED-TO-KNOW-RISKS yes). If this patient encounters any unexplained new words, its need to know the definition triggers asking the question "What's that?" Similarly, if it needs to know any property values it does not yet know, it asks an associated, standard question, such as "Are there any risks?" At the moment, these are modeled as reflexes that do not invoke planning or reasoning (cf. the Reflexive action path in figure 2.1). Naturally, these processes could be expanded into full-fledged perception-reasoning-action cycles: the agent could detect its need to know something, create a plan to learn it, and instantiate that plan. There is nothing wrong with this more expansive approach except that it takes more time to implement, and it is questionable whether it models what people actually do. Our point here is to show various knowledge engineering options that can be mixed and matched in actual agent-development efforts.

In response to the first question ("What's that?"), the doctor might answer "It's an endoscopic procedure." Assuming the patient already knows what endoscopy is, it will generate the following TMR:

EGD-1

    IS-A   ENDOSCOPY-1

Then, when it proceeds to ontology learning—described in section 7.2.2—it will record this information as generic ontological knowledge.

When the agent then asks "Is it risky?", the doctor might respond "Not very." One lexical sense of the construction *Not very* (which is formally similar to yes-adv2, presented earlier) anticipates its use as a fragmentary utterance and records its meaning as "RISK .3" The agent learns this as ontological knowledge as well.

Having gathered all the needed information, the agent moves on to the other subevents in the RESPOND-TO-PROPOSED-MEDICAL-INTERVENTION script, namely:

- RUN-EVALUATION-FUNCTION, which is a mathematical function that applies different weights to different property values and then combines them into a decision to agree to the procedure or not.

- FORMULATE-RESPONSE, which involves selecting which details to convey, how polite to be, and so on. This results in an MMR that is passed to the Action Specification module in the normal way.

To conclude, this chapter has explained an agent's participation in dialog applications as an example of its standard perception-reasoning-action cycle. We have emphasized the balance of responsibility across knowledge bases and programs, and how our knowledge engineering methodology makes agent operation explainable by agents and developers, as well as extensible over time.

7 Learning

A core capability of LEIAs is lifelong, humanlike learning, both through reading and through multimodal interactions with people. This chapter describes how LEIAs learn new words, phrases, ontological concepts, and properties of concepts, and it illustrates this learning using examples from application systems.[1] The chapter is divided into two parts that can be read in either order or in parallel.

- Part 1 is illustrative: it describes four modes of learning using examples from implemented demonstration systems and serves as a soft introduction to the kinds of reasoning that learning agents must be capable of.

- Part 2 is conceptual: it details the eventualities a LEIA can encounter while learning lexicon and ontology and presents the scaffolding into which all of the phenomena illustrated in part 1 neatly fit.

Consulting the table of contents while reading this chapter might be helpful for linking the examples to the conceptual scaffolding.

While human-level competency in learning remains a long-term goal, the microtheory of learning, like all LEIA microtheories, is geared toward equipping agents with useful capabilities as soon as possible.

A terminological note: *Candidate knowledge* refers to information that has been analyzed by a LEIA but has not yet been added to its lexicon or ontology.

7.1 Part 1: An Example-Based Introduction to Different Modes of Learning

This section describes LEIA learning in terms of the following four modes, with "modes" being used informally to capture key distinctions.

- **Basic learning through language**: a language input offers learnable information that the agent understands with high confidence and can acquire without the need for additional evidence (section 7.1.1).
- **Mixed-initiative learning**: the agent supplements basic learning by initiating dialog with a human partner to ask for clarification or request additional information (section 7.1.2).
- **Data-driven learning**: the agent leverages data-driven methods to collect additional material to support specific learning goals and to reason over the combined evidence. Data-driven learning relies on basic learning for the analysis of the language inputs selected for the task (section 7.1.3).
- **Multimodal learning**: multiple channels of input—such as language and vision—inform the learning process (section 7.1.4). Multimodal learning can be mixed-initiative and/or include data-driven methods.

These are not the only modes of learning available to LEIAs: they can also learn through nonlinguistic channels of perception and through introspection—for example, when engaging in memory management. These language-free modes of learning are as important to LEIA modeling as the others, but they are not a priority here due to this book's emphasis on enabling explanation through language.

Part 1 of this chapter has four objectives:

1. to illustrate the workflows of various modes of learning;
2. to show how a LEIA's generic learning capabilities apply across domains and applications;
3. to underscore that deep language understanding is needed for learning-oriented reasoning; and
4. to distinguish between generic learning capabilities and the ways in which they play out in particular applications.

7.1.1   Basic Learning through Language
Figure 7.1 shows the basic flow of lexicon and ontology learning from a language input.

The analysis procedures and decision-making functions referred to in the diagram are modeled using concepts recorded in the AGENT-LEARNING subtree of the ontology.

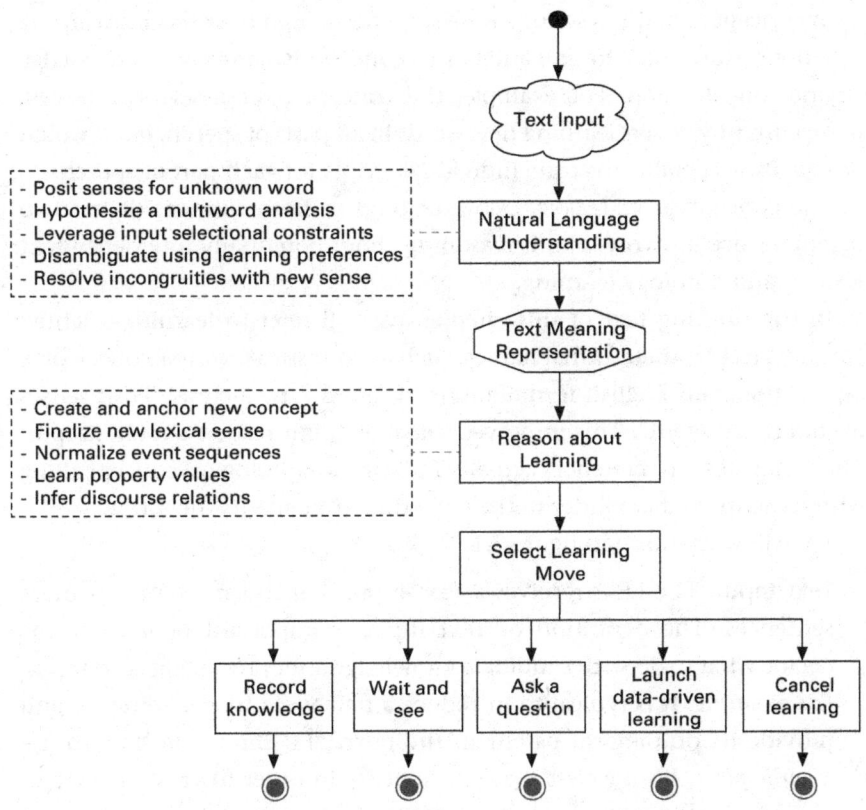

Figure 7.1
The basic flow of learning lexicon and ontology through language.

```
AGENT-LEARNING
    LEXICON-LEARNING-FUNCTIONS
        POSIT-SENSES-FOR-UNKNOWN-WORD
        HYPOTHESIZE-MULTIWORD-ANALYSIS
        LEVERAGE-INPUT-SELECTIONAL-CONSTRAINTS
        DISAMBIGUATE-USING-LEARNING-PREFERENCES
        RESOLVE-INCONGRUITIES-VIA-NEW-SENSE
        FINALIZE-NEW-LEX-SENSE
        SELECT-LEARNING-MOVE
    ONTOLOGY-LEARNING-FUNCTIONS
        CREATE-AND-ANCHOR-NEW-CONCEPT
        NORMALIZE-EVENT-SEQUENCES
        LEARN-PROPERTY-VALUES
        INFER-DISCOURSE-RELATIONS
        SELECT-LEARNING-MOVE
```

Each concept contains a script whose steps detect and treat specific learning situations. How much to split and bunch concepts is, as always, a knowledge engineering decision. For example, the concept POSIT-SENSES-FOR-UNKNOWN-WORD currently covers learning new words in all parts of speech, but it would be equally acceptable to create individual scripts for each part of speech.

The concept SELECT-LEARNING-MOVE belongs to both subtrees (it has two parents) since it involves joint reasoning about potentially interdependent lexicon and ontology learning.

In the running text of this chapter, we will refer to learning-oriented concepts not by their formal names, such as POSIT-SENSES-FOR-UNKNOWN-WORD, but by the plain English formulations in figure 7.1., such as "Posit senses for unknown word." This improves readability, but readers should keep in mind that all such references actually indicate an ontological concept along with its script and a pointer to the code that implements the script.

We will work through figure 7.1.

- **Text Input.** The LEIA receives a Text Input, which can be one or more sentences. The definition of Text Input is important because agents decide what to do with candidate knowledge after processing each input. For example, it is typical to introduce a new word in one sentence and provide its ontological parent in the next: "I'll show you how to use a *garlic press*. It's *a gadget* that you squeeze in order to crush a clove of garlic." If the agent tries to learn the meaning of *garlic press* after just the first sentence, it will not know where to attach the new concept in the ontological hierarchy, whereas if it waits for the next sentence, the ontological parent becomes clear. What constitutes an input must be determined within an application. In data-driven learning, developers will select input length or design a function to determine it. In text-based dialog applications with strict turn taking, an input is the text submitted as one turn. In spoken dialog applications or text-based dialog applications without strict turn-taking, the agent needs to dynamically determine what constitutes an input—for example, based on pauses in the interlocutor's speech or typing.

- **Natural Language Understanding.** The agent analyzes the Text Input using its Natural Language Understanding engine. This process includes learning procedures that are triggered when the agent encounters unknown words and phrases. These procedures include (see the upper-left box in figure 7.1) positing senses for unknown words, hypothesizing multiword expression analyses, using selectional constraints to infer the meaning of unknown words, using learning-oriented preferences to disambiguate elements of input, and resolving incongruities in meaning representations by creating new senses of known words.

- **Text Meaning Representation.** The output of Natural Language Understanding is a Text Meaning Representation (TMR) that reflects the results of learning-oriented reasoning. The TMR includes (a) the most specific semantic analysis of newly learned words and phrases that the agent is able to formulate using the methods available so far, and (b) semantic analyses of whatever new world knowledge might have been presented, which will be incorporated into the ontology during ontology learning. For example, if a language input is *Endoscopy is a very low-risk procedure*, the TMR will represent that meaning as "ENDOSCOPY (RISK .1)." During ontology learning, the agent will use this meaning representation to add the property value "RISK .1" to the concept ENDOSCOPY.

- **Reason about Learning.** Most scripts in this module involve reasoning in service of learning ontological concepts and their properties. However, certain lexicon learning procedures are carried out as well, such as remapping tentative new word senses to newly learned ontological concepts.

- **Select Learning Move.** Finally, the agent must decide what to do with the candidate knowledge. There are five options. The first one, record the knowledge, is selected in optimal cases, when the candidate knowledge is of high quality and high confidence. In suboptimal cases—when something is underspecified or missing—the agent chooses among the other four options: wait and see if the needed information is forthcoming; ask a human a question about it; launch data-driven learning to try to find the needed information; or cancel the learning of the entity, deeming it unnecessary within the given application.

Below is an example of an input that the LEIA can process well enough to directly record the candidate knowledge in its lexicon and ontology. In this example, the LEIA is being trained in the medical domain by a human who explains:

Systemic sclerosis is a multisystemic autoimmune disease of unknown origin that affects connective tissue. It often affects the esophagus and can cause heartburn, regurgitation, dysphagia, and chest pain.

To understand how the agent processes this input, it is important to understand the initial state of its lexicon. It knows all of the words except for *systemic* and *sclerosis*; it knows the multiword expressions *autoimmune disease, of unknown origin*, and *connective tissue*; and it has a construction that anticipates nominal compounds of the form "Noun_BODY-PART *pain*," which enables the precise analysis of *chest pain* as "PAIN (LOCATION CHEST)."

Of the ten learning functions shown in figure 7.1—five attached to Natural Language Understanding and five attached to Reason about Learning—six are leveraged in processing this input:

- Posit senses for unknown word: the agent posits senses for the adjective *systemic* and the noun *sclerosis*.

- Hypothesize a multiword analysis: the agent reasons that these words likely form the multiword expression *systemic sclerosis*.

- Disambiguate using learning preferences: the agent disambiguates *is* using a definitional sense that maps to the ontological relation IS-A, and it disambiguates *the* in *the esophagus* by preferring a generic interpretation due to the lack of a coreferential antecedent and the fact that ontology learning is underway.

- Create and anchor new concept: the agent generates the concept SYSTEMIC-SCLEROSIS from the newly learned lexical sense *systemic sclerosis*, and it anchors the new concept in the ontology as a child of AUTOIMMUNE-DISEASE based on the previously established IS-A relation.

- Finalize new lexical sense: The agent remaps the lexical sense *systemic sclerosis* from EVENT, which was the initial mapping during TMR creation, to the new concept SYSTEMIC-SCLEROSIS.

- Learn property values: The agent learns all of the property values presented in the input, as shown in the final concept frame it learns.

SYSTEMIC-SCLEROSIS		; systemic sclerosis
IS-A	AUTOIMMUNE-DISEASE	; is an autoimmune disease
THEME	SET-1	; that affects multiple systems
CAUSED-BY	IDIOPATHIC-EVENT	; it is of unknown origin
EFFECT	CHANGE-EVENT-1	; it affects connective tissue
EFFECT	CHANGE-EVENT-2	; it often affects the esophagus
HAS-TYPICAL-SYMPTOM	SET-2	; it has a set of typical symptoms
SET-1		
MEMBER-TYPE	ANATOMICAL-STRUCTURE	
CARDINALITY	>1	

CHANGE-EVENT-1
    THEME   CONNECTIVE-TISSUE
CHANGE-EVENT-2
    THEME   ESOPHAGUS
    FREQ-ACROSS-INSTANCES   .7   ; "often"
SET-2
    ELEMENTS   DYSPHAGIA, PAIN-1, REGURGITATE, HEARTBURN
PAIN-1
    LOCATION   CHEST

This frame looks very similar to the TMR that was produced as a result of language understanding. The key differences are: numerical indices are used differently in TMRs and ontological frames; and whereas the TMR mapped *systemic sclerosis* to the generic EVENT since the agent had not yet learned the concept SYSTEMIC-SCLEROSIS, the ontological frame uses the newly learned concept.

The final stage of learning is Select Learning Move, where the agent evaluates whether its candidate additions to the lexicon and ontology are fit to be recorded to memory. In this example, all of them are.

To reiterate the point of this example: in some cases, language inputs are sufficient to seed learning without the agent needing to ask a human for clarification or consult a text corpus for additional information. It behooves teachers of LEIAs to present information in a clear and organized way, just as teachers strive to do when interacting with human students.

Of course, it will happen that the agent's analysis of an input is not sufficient to support confident learning: there might be residual ambiguities in the language input, a new term might not have been defined, and so on. In dialog applications, the agent can then ask the human for help. This is called mixed-initiative learning, to which we now turn.

7.1.2 Mixed-Initiative Learning

Mixed-initiative learning is dialog-based learning in which both the teacher and the trainee can initiate moves. In mixed-initiative learning, the big questions are *when*, *why*, and *how* an agent takes the initiative to ask questions. This decision-making is informed by two kinds of parameters: those involving learning per se and those relevant to specific applications.

- **Parameters involving learning per se.** The agent knows basic principles about the minimal requirements for learning lexical and ontological knowledge and, therefore, can detect when candidate knowledge is not yet learnable. For example, if the agent is only able to map a new word to a very generic ontological concept, such as OBJECT or EVENT, it knows that it needs to identify a more precise mapping before learning the lexicon entry for the word. Similarly, if the agent is trying to learn a concept but does not know where to attach it in the ontological hierarchy, it knows that it needs to identify the appropriate parent. The agent detects such needs and places the goal of fulfilling them on its agenda as it works through the reasoning functions in figure 7.1.

- **Parameters relevant to specific applications.** In different applications, agents can have different needs, preferences, and priorities. For example, one virtual patient might be very concerned about the pain associated with medical procedures, and another might be concerned about the risks. These special interests—the need-to-know features described earlier—cause patients to seek out different kinds of information before making a decision about a recommended medical procedure. Need-to-know features are modeled as standing goals that are placed on the agent's agenda when it is instantiated in an application run. For every input that concerns these features, the agent must decide whether to immediately ask a question or wait and see if the information is forthcoming.

We will illustrate mixed-initiative learning using a run of the Maryland Virtual Patient prototype clinician training system, which allows human trainees to diagnose and treat virtual patients in open-ended simulations. The following is an excerpt from a system run that includes material to be learned presented by the doctor (D) and follow-up questions asked by the virtual patient (P).

The agent learns new vocabulary, associated concepts, and property values in the same way as was illustrated using the systemic sclerosis example above. In addition, it decides how to respond to a proposed procedure using its decision function AGREE-TO-AN-INTERVENTION-OR-NOT, which is part of its ontological knowledge about clinical medicine.[2] Details aside, what is important for this discussion of mixed-initiative learning is how and why the agent decides to ask questions.

As already noted, in the Maryland Virtual Patient system, all virtual patients are provided with an inventory of personality traits that complements their medically oriented property values. This allows for the creation of a large, highly differentiated population of virtual patients that present

Table 7.1
Learning lexicon and ontology through a mixed-initiative dialog interaction.

Dialog	Ontological knowledge learned	Lexical knowledge learned
D: You have achalasia.	The concept ACHALASIA is learned and made a child of DISEASE.	The noun *achalasia* is learned and mapped to the concept ACHALASIA.
P: Is it treatable? D: Yes.	The value for the property TREATABLE in the ontological frame for ACHALASIA is set to *yes*.	
D: I think you should have a Heller myotomy.	The concept HELLER-MYOTOMY is learned and made a child of MEDICAL-PROCEDURE. Its property TREATMENT-OPTION-FOR receives the filler ACHALASIA.	The noun *Heller myotomy* is learned and mapped to the concept HELLER-MYOTOMY.
P: What is that? D: It is a type of esophageal surgery.	The concept HELLER-MYOTOMY is moved in the ontology tree: it is made a child of SURGICAL-PROCEDURE. Also, the THEME of HELLER-MYOTOMY is specified as ESOPHAGUS.	
P: Are there any other options? D: Yes, you could have a pneumatic dilation instead, which is an endoscopic procedure.	The concept PNEUMATIC-DILATION is learned and made a child of ENDOSCOPY.	The noun *pneumatic dilation* is learned and mapped to the concept PNEUMATIC-DILATION.
P: Does it hurt? D: Not much.	The value of the property PAIN-LEVEL in PNEUMATIC-DILATION is set to .2 (on the scale $\{0,1\}$).	

different clinical challenges to system users. Among other things, patients' personality traits make them interested in knowing certain information relevant to their health care that informs their decision-making. The patient in the scenario above wants to know, for every diagnosed medical condition, whether it is treatable, and for every proposed procedure, what it is and how painful it is. After each of the doctor's utterances, the patient needs to decide what to do, and its decisions are informed by whether any of its standing goals has been triggered. If so, it decides whether to immediately act on it or wait and see. This decision is determined by a function whose features include the patient's personality trait called EAGERNESS-TO-KNOW and the recency of the patient's last question in the dialog overall.

To summarize this section, mixed-initiative learning, like all language-based learning, follows the algorithm in figure 7.1. A patient asks a question when a standing goal involving a learning need is triggered. That goal can involve basic needs of learning, such as where to attach a learned concept in the ontology, or specifics of a particular agent within a particular application, such as an agent's need to know whether a proposed procedure is painful. When a standing goal has been triggered, the agent has to decide what to do about it. In the case of a dialog application, the typical options are wait and see or ask a question.

7.1.3 Data-Driven Learning

An agent can choose to pursue data-driven (corpus-based) learning either as a result of decision-making after the basic learning process in figure 7.1 ("Launch data-driven learning") or because it is tasked with learning by reading. Figure 7.2 shows the workflow for data-driven learning, which is the same as for basic learning but with the additional preparation of learning material.

Although data-driven learning is, by its nature, not explainable, the decision-making around it is, and the associated knowledge is recorded in the ontology under the concept "Corpus curation for learning." For example, there must be methods for selecting a corpus, ordering what is to be learned, deciding which kinds of additional data are needed to support the learning of a given entity, and so on. Even if some of these decisions are taken by human developers, they still involve knowledge, and that knowledge needs to be part of an agent's world model.

Consider some situations in which data-driven learning is useful.

- The agent is learning a new word and associated concept, but it cannot determine where the concept belongs in the ontological hierarchy. For this, it would be useful to find a definitional sentence such as *Polo is a sport played while riding horses.*

- The agent encounters a word in a context that gives insufficient clues to its meaning. For example, *She loves khachapuri* is far less useful for learning than *She eats khachapuri all the time*. From the latter, the agent can learn that *khachapuri* is a kind of FOOD, which is a good starting point for further investigation of the properties of this Georgian stuffed bread.

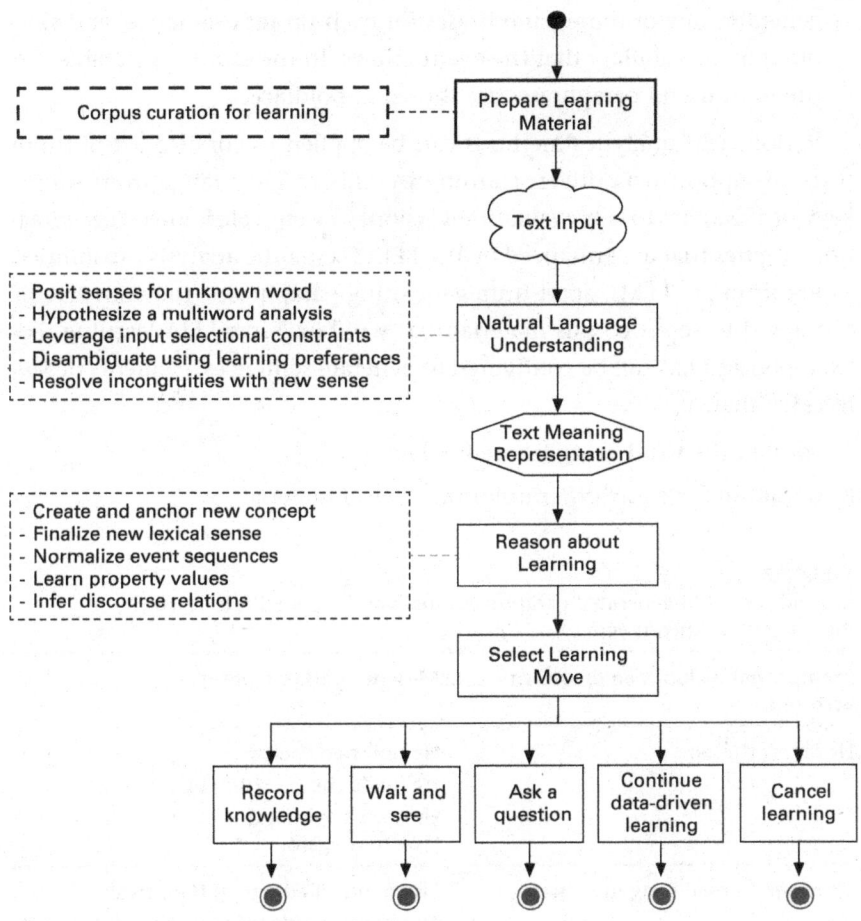

Figure 7.2
Lexicon and ontology learning by a LEIA informed by data-driven methods.

- The agent has learned a new concept but does not yet know any of the property values that differentiate it from its parent. For this, examples that include distinguishing property values would be useful.
- The agent is trying to learn a new verb, which is difficult in principle because a verb's meaning cannot be inferred from the meanings of its arguments. For example, one cannot guess what *buff* means from the sentence *He buffed the car.* For new verb learning, paraphrasing is useful, and large language models (LLMs) are well suited to the task. Our lab has developed an LLM-based paraphrasing tool that serves this purpose.[3] It

generates one or more paraphrases for each prompt sentence, and they often use vocabulary that the agent knows. In the examples in table 7.2, the most useful paraphrases are shown in boldface.

Various text analytics methods can be applied to curate a set of input texts to support data-driven learning by LEIAs.[4] They range from simple keyword searches to lazy evaluation versions of web-scale knowledge extraction engines that are enhanced by the LEIA's semantic analysis capabilities.

Returning to LLMs, apart from generating paraphrases, they can also be prompted to select corpus material that will best serve LEIA learning. For example, an LLM can be configured to generate sentences, elements of *SentenceSet*, that:

1. contain the word being learned: *Word*;

2. do not include any other unknown lexical material;

Table 7.2
Examples of LLM-generated paraphrases that can help a LEIA to learn the meaning of unknown verbs.

Prompt that includes an unknown verb sense	LLM-generated paraphrases
He *buffed* the car.	– **He polished the car.** – He made the car shiny by rubbing it. – He used a special cloth to make the car look new again.
They *offed* a rival gang member.	– **The gang killed one of their rivals.** – **The gang murdered someone who was in a different gang.** – The gang eliminated a member of a rival group. – A member of a rival gang was killed by the gang.
This cake *cuts* easily.	– **This cake is easy to cut.**
They *settled on* what to do.	– **They agreed on a decision.** – They made a choice about what to do. – They came to a conclusion on what actions to take. – They reached a resolution on the course of action. – **They decided on a plan of action.**

3. include, among others, definitional sentences for *Word*; TMRs for these sentences will help to determine where in the ontological hierarchy to anchor the concept (*Concept*) that reflects the meaning of *Word*; and

4. include, among others, descriptive sentences specifying constraints on the values of properties defined for *Concept*; the set of such properties will be inherited from *Concept*'s ancestors in the ontological hierarchy; a prerequisite is to engineer prompts to the LLM for each property in the ontology.

SentenceSet is also expected to include sentences that do not contain either a definition or properties that the agent knows about. Such sentences could trigger the learning of new properties, but this option must be invoked judiciously (see the discussion of property creation in section 7.2.2).

Word may be polysemous, and some of its senses may already be recorded in the LEIA's lexicon. If filtering out known senses of *Word* is necessary, then the following process must be carried out:

1. Prompt the LLM with *Word* in all its *WordSenses*: for example, "Generate sentences / phrases with *Word* in multiple senses."

2. Pass all sentences through a transformer encoder, extract the contextualized embeddings for all instances of *Word*, and cluster them in this multidimensional space.

3. Generate TMRs for a sampling of sentences from each cluster in order to understand which cluster corresponds to which known sense of *Word*. Filter those out of *SentenceSet* and concentrate on the cluster corresponding to the unknown sense of *Word*.

4. Posit a concept to reflect the unknown sense of *Word*.

5. Anchor that concept in the ontological hierarchy. If any of the sample sentences included a definition, then use the IS-A link in its TMR to anchor the new concept; otherwise, generate a definitional sentence—such as "What does *Word* mean in *Sentence*?" (with *Sentence* being selected from the appropriate cluster)—and use the LLM's definitional response to anchor the concept in the hierarchy.

This learning process is sure to meet with complications. For example:

• Learning is recursive: in order to learn a given word or concept, the system might first need to learn another one.

- Although the system might know *some* senses of all but one word in the sample sentences, the senses it knows might not be the ones needed in the context. This will lead to a semantic incongruity in the TMR and make it less than optimal for learning the new word.
- Word sense clustering can yield multiple clusters, which may trigger the need to learn more than one new word sense.
- Current LLMs can generate incomplete or fallacious outputs. This makes it prudent to integrate the use of LLMs with old-style data analytics and human interaction to support automatic learning.

The approach to learning we have described differs in principle from recent work on lifelong learning for textual understanding[5] in that, for us, learning is an integral part of the overall continuous performance of a knowledge intensive system. Note also that the terms *lifelong learning* and *continual learning*, when used in the context of neural networks, refer exclusively to designing networks capable of effectively learning a new classification task without entirely losing the ability to do a previous classification task,[6] which is a far more restricted usage of the term than ours.

The application of LLMs we have described introduces a novel view of neurosymbolic architectures. Traditionally, neurosymbolic approaches to AI attempt to reincorporate the structure and speed of symbolic reasoning into the flexible representations of deep learning. We present an inverse approach in which LLMs support learning that is based on symbolic ontological and linguistic knowledge.

7.1.4 Multimodal Learning

LEIAs treat inputs from all channels of perception in the same way: they translate them into ontologically grounded meaning representations (XMRs) and then reason over those representations. Accordingly, learning-oriented reasoning is the same no matter the channel of perception that encountered the learnable information. This section presents an example of a LEIA embedded in a furniture-building robot, which illustrates multimodal learning involving inputs from vision and language.[7]

Robots typically have some inventory of physical actions they can perform, some inventory of objects they can recognize and manipulate, and, in some cases, rudimentary language skills—such as being able to react to certain vocal commands.[8] A LEIA-robot hybrid, by contrast, can acquire a mental model of these actions and objects through dialog with human collaborators. That is, people can help embodied robotic LEIAs to understand their world by naming objects and actions; describing actions in terms of their causal organization, prerequisites, and constraints; listing the affordances of objects; and explaining people's expectations of the robots. This kind of understanding will enhance LEIAs' ability to understand their own actions and the actions of others, and to become more humanlike collaborators overall. Clearly, this kind of learning relies on semantically interpreting language inputs, and it mirrors a major mode of learning in humans—learning through language. In this section, we describe our work on integrating a LEIA with a robot in an application system.

The system we describe is a social robot collaborating with a human user to learn complex actions. The experimental domain is the familiar task of furniture assembly that is widely accepted as useful for demonstrating human-robot collaboration on a joint activity. Roncone, Mangin, and Scassellati (2017) report on a Baxter robot supplied with high-level specifications of procedures that implement chair-building tasks, which are represented in the hierarchical task network (HTN) formalism.[9] In that system, the robot uses a rudimentary sublanguage to communicate with the human in order to convert these HTN representations into low-level task planners capable of being directly executed by the robot. Since the robot does not have the language understanding capabilities or the ontological knowledge substrate of LEIAs, it cannot learn by being told or reason explicitly about the HTN-represented tasks. As a result, those tasks have the status of uninterpreted skills stored in the robot's procedural memory.

We developed a LEIA-robot hybrid based on the robot just described. The resulting system was able to

- learn the meaning of the initially uninterpreted basic actions;
- learn the meaning of operations performed by the robot's human collaborator from natural language descriptions of them;
- learn, name, and reason about meaningful groupings and sequences of actions;
- organize those sequences of actions hierarchically; and
- integrate the results of learning with knowledge stored in the robotic LEIA's semantic and episodic memories.

To make clear how all this happens, we must start from the beginning. The robotic LEIA brings to the learning process the functionalities of both the LEIA and the robot. Its robotic side can (a) visually recognize parts of the future chair (e.g., the seat) and the tools to be used (e.g., a screwdriver) and (b) perform basic programmed actions, which are issued as non-natural-language commands such as GET(LEFT-BRACKET), HOLD(SCREWDRIVER), and RELEASE(LEFT-BRACKET). The hybrid system's LEIA side, for its part, can generate ontologically grounded meaning representations (XMRs) from both user utterances and physical actions. The interactive learning process that combines these capabilities is implemented in three modules.

Learning module 1: Concept grounding   The robotic LEIA learns the connection between its basic programmed actions and the meaning representations of utterances that describe them. This is done by the user verbally describing a basic programmed action at the same time as launching it. For example, the user can say, *You are fetching a screwdriver* while launching the procedure GET(SCREWDRIVER). The robotic LEIA generates the following TMR while physically retrieving the screwdriver. Fetching is interpreted as changing the location of the object such that it ends up beside the individual who asked the robot to fetch it.

CHANGE-LOCATION-1
    THEME   SCREWDRIVER-1
    AGENT   ROBOT-1
    TIME   *find-anchor-time*
    EFFECT   BESIDE-1
    textstring   fetch
    lex-sense   get-v1   ; fetch is recorded as a synonym of get
BESIDE-1
    DOMAIN   SCREWDRIVER-1
    RANGE   HUMAN-1   ; the human asking for it to be fetched

Learning module 2: Learning legal sequences of known basic actions   The robot learns legal sequences of known basic actions by hierarchically organizing the TMRs for sequential event descriptions. It recognizes these sequences as new complex actions (ontological scripts), which it names and records in its ontology. We illustrate the long process of chair assembly by tracing how the robot learns how to assemble the third of the four chair legs (see Nirenburg & Wood, 2017 for details).

- The user says, "We are building the right back leg."
- The robotic LEIA carries out a mental action: it generates a TMR for that utterance.
- The user says, "Get another foot bracket" and launches the associated robotic action by inputting GET(BRACKET-FOOT).
- The robotic LEIA carries out a sequence of physical actions: first, it undertakes the asserted GET(BRACKET-FOOT) action. Then it carries out the action it typically performs next: RELEASE(BRACKET-FOOT). The robotic LEIA also carries out a mental action: it learns to associate this complex event with the TMR for "Get another foot bracket."
- The user says, "Get the right back bracket" and launches the associated robotic action by inputting GET(BRACKET-BACK-RIGHT).
- *The robotic LEIA performs the associated physical and learning actions, as before.*
- The user says, "Get and hold another dowel" and launches the associated robotic actions by inputting GET(DOWEL), HOLD(DOWEL).
- *The robotic LEIA performs the associated physical and learning actions.*
- The user says, "I am mounting the third set of brackets on a dowel."
- The robotic LEIA carries out a mental action: it generates a meaning representation of this utterance.
- The user carries out a physical action: he affixes the foot and the right back brackets to the dowel.
- The robotic LEIA learns through demonstration: it observes this physical action and generates a meaning representation of it.
- The user says, "Finished."
- The robotic LEIA carries out a mental action: it generates a meaning representation of this utterance.
- The user says, "Release the dowel" and launches the associated robotic action by inputting RELEASE(DOWEL).
- *The robotic LEIA performs the associated physical and learning actions.*
- The user says, "Done assembling the right back leg."
- The robotic LEIA carries out a sequence of mental actions:
    a. It generates a meaning representation for that utterance.
    b. It learns the action subsequence for ASSEMBLE-RIGHT-BACK-LEG.

c. It learns the following ontological concepts in their meronymic relationship:

RIGHT-BACK-LEG

    HAS-OBJECT-AS-PART   BRACKET-FOOT, BRACKET-BACK-RIGHT, DOWEL

d. It learns that RIGHT-BACK-LEG fills the HAS-OBJECT-AS-PART slot of CHAIR.

Learning module 3: Memory management for newly acquired knowledge Newly learned process sequences, such as ASSEMBLE-RIGHT-BACK-LEG, and newly learned objects, such as RIGHT-BACK-LEG, must be incorporated into the robotic LEIA's long-term semantic and episodic memories. For each newly learned concept, the memory management module first determines whether this concept should be added to the LEIA's semantic memory or merged with an existing concept. To make this choice, the agent uses an extension of the concept-matching algorithm reported in English and Nirenburg (2007) and Nirenburg, Oates, and English (2007). This algorithm is based on unification, with the added facility for naming concepts and determining their best position in the ontological hierarchy. Details aside, the matching algorithm works down through the ontological hierarchy—starting at the PHYSICAL-OBJECT or PHYSICAL-EVENT node, as applicable—and identifies the closest match that does not violate any recorded constraints. Nirenburg, Oates, and English (2007) describe the eventualities that this process can encounter.

To recapitulate, the system described here concentrates on a robotic LEIA learning through language understanding in a context that involves visual perception as well. These capabilities to allow the robot to (a) perform complex actions without the user having to spell out a complete sequence of basic and complex actions; (b) reason about task allocation between itself and the human user; and (c) test and verify its knowledge through dialog with the user, avoiding the need for the large number of training examples that might be required when learning through demonstration alone.

The work on integrating linguistically sophisticated cognitive agents with physical robots offers several advantages. First, robotic LEIAs can explain their decisions and actions in human terms, using natural language. Second, their operation does not depend on the availability of big-data training materials; instead, we model the way people learn, which is largely through natural language interactions. Third, our work overtly models the robotic LEIA's memory components, which include the implicit memory of skills (the robotic component), the explicit memory of concepts (objects, events, and their properties), and the explicit memory of concept instances, including episodes, which are represented in our system as hierarchical transition networks. The link between the implicit and explicit layers of memory allows the robot to reason about its own actions.[10]

There are several natural extensions to this work. After the robot's physical actions are grounded in ontological concepts, the robot should be able to carry out commands or learn new action sequences by acting directly on the user's utterances, without the need for direct triggering of those physical actions through software function calls. In addition, the incorporation of text generation and dialog management capabilities—which were not included in the reported work but are currently available—would both allow the robot to take a more active role in the learning process, as by asking questions, and enrich the verisimilitude of interactions with humans during joint task performance.

7.1.5   An Extended Example: A LEIA Learns Rules of the Road

The previous four sections have presented short examples that illustrated different modes of learning. This section presents a more extended example that illustrates the important interaction between language understanding, reasoning about language, and reasoning about learning. The description is somewhat technical, but these details are needed to show that LEIAs operate using a generalized workflow that leverages knowledge and reasoning that are as generic as possible but as specific as needed, thus emulating human knowledge and reasoning. The application is simulated driving, and the scenario involves a human teaching a LEIA what to do at a four-way stop.[11]

Human: I'll teach you what to do at a four-way stop.
A four-way stop is a four-way intersection where all vehicles must come to a stop.
You can recognize a four-way stop because there is a stop sign on each of the four corners.
A stop sign is an octagonal red sign with white letters.
It is usually about 30 inches across.
But it can also be 35 inches wide.
After you stop at a four-way stop, you must decide when to start driving.
If there are no other vehicles at the intersection, you can go.

LEIA: How do I tell if a vehicle is at the intersection?

Human: A vehicle is at the intersection if it is adjacent to a stop sign and not moving.

LEIA: I think a vehicle is beside a stop sign when it is no more than 50 inches away. Is that right?

Human: Yes.

If other vehicles arrived at the intersection before you, they have the right of way.

Before the learning started, the LEIA's lexicon included the necessary meanings of many of the individual words as well as the meanings of the constructions *I'll teach you what to do, a NP is a NP* (indicating the IS-A relation), *four-way intersection, come to a stop, how do I tell if, no more than*, and *have the right of way*.

Human: I'll teach you what to do at a four-way stop. The agent analyzes this sentence using the lexical sense for the construction *I'll teach you what to do PP*[12]. The basic semantic interpretation of that lexical sense, which is recorded in its sem-struc, says:

- This is an instance of TEACH in which the speaker is the AGENT and the listener is the BENEFICIARY.

- The THEME of TEACH is the events that comprise some script, which is represented as EVENT.SUBEVENTS.

- That EVENT has some RELATION to the meaning of the object of the preposition—here, *a four-way stop*.

When this lexical sense is used to analyze our example, it generates the following TMR.

TEACH-1
AGENT	HUMAN-1	; I am teaching
BENEFICIARY	LEIA-1	; you
THEME	EVENT-1.SUBEVENTS	; a new script (an EVENT with SUBEVENTS)
HAS-MP	CREATE-AND-ANCHOR-NEW-CONCEPT: OBJECT-1	
	CREATE-AND-ANCHOR-NEW-CONCEPT: EVENT-1	

EVENT-1   ; the new EVENT "what to do at a four-way stop"
RELATION	OBJECT-1	; is related to the new object "a four-way stop"

Some notes about this TMR:

- The meaning of *a four-way stop* is underspecified—that is, it is mapped to OBJECT—because this is just the initial stage of learning the new expression.

- The meaning procedures (fillers of HAS-MP) serve as input to the ontology-learning function CREATE-AND-ANCHOR-NEW-CONCEPT; they indicate which language strings need to be converted into concepts and then incorporated into the ontology. Here, "four-way stop" will be a newly learned OBJECT and "what to do at a four-way stop" will be a newly learned EVENT.

The result of running these meaning procedures is the agent learning the following knowledge:

lexicon: the nominal sense stop-n1, which requires the modifier *four-way* and is mapped to the concept FOUR-WAY-STOP

ontology: the concept FOUR-WAY-STOP, which becomes a child of OBJECT; and the concept EVENT-AT-A-FOUR-WAY-STOP, which becomes a child of EVENT

At this point in the process, the question is: What constitutes the first input in this dialog? Recall that the agent decides what to do with candidate knowledge only after each input, which is not necessarily a single sentence. For this scenario, we assume that the agent considers each sentence a new input, but its default decision about what to do next is wait and see. It overrides that decision if it has a burning question (illustrated below) or when learning about the given topic concludes—which may be indicated, for example, by the human asking if everything is clear.

Human: A four-way stop is a four-way intersection where all vehicles must come to a stop. The first part of this sentence indicates that the ontological parent of FOUR-WAY-STOP is FOUR-WAY-INTERSECTION. Before this learning scenario, the agent already knew the multiword expression *four-way intersection*, mapped to FOUR-WAY-INTERSECTION, and it understands the subsumption information thanks to a lexical sense that maps the construction *a Noun is a Noun* to the property IS-A. This interpretation of the polysemous word *is* is preferred thanks to a heuristic in "Disambiguate using learning preferences" that prefers subsumption analyses when the agent is in learning mode. So, the TMR for the first part of the sentence is:

FOUR-WAY-STOP-1
    IS-A   FOUR-WAY-INTERSECTION-1

This information triggers a function in "Create and anchor new concept" that changes the parent of FOUR-WAY-STOP from OBJECT to FOUR-WAY-INTERSECTION.

The second part of this sentence, *where all vehicles must come to a stop,* expresses certain property values of a FOUR-WAY-STOP:

FOUR-WAY-STOP-1		; a four-way stop
IS-A	FOUR-WAY-INTERSECTION-1	; is a four-way intersection
LOCATION-OF	MODALITY-1	; where
MODALITY-1		
TYPE	OBLIGATIVE	
VALUE	1	; it is obligatory
SCOPE	STOP-DRIVING-EVENT-1	; to stop driving
STOP-DRIVING-EVENT-1		
THEME	SET-1	
SET-1		; all vehicles
MEMBER-TYPE	VEHICLE	
QUANT	1	

When the agent evaluates this TMR for learnable information, specific ontology-learning functions included in "Learn property values" facilitate appropriately attaching the property values in the TMR to the concept description of FOUR-WAY-STOP.

Human: You can recognize a four-way stop because it has a stop sign on each of the four corners. The lexicon includes the construction *You can recognize DirectObj because Clause.*[13] Informally, the semantic interpretation is as follows (remember that ^ indicates "the meaning of").

you can	recognize	DirectObj	because	Clause
MODALITY(POTENTIAL 1)	DETECT	^DirectObj	HAS-SALIENT-FEATURE	^Clause

Before considering how this applies to our reasoning-heavy example, let us consider how this construction plays out for a simpler example: *You can recognize a pug because it has a short snout*:

MODALITY-1	
TYPE	POTENTIAL
VALUE	1
SCOPE	DETECT-1
DETECT-1	
THEME	PUG-DOG-1
PUG-DOG-1	
HAS-SALIENT-FEATURE	SNOUT-1
SNOUT-1	
LENGTH	.2

This says that it is possible to recognize a pug because of the salient feature of a short snout. Our example—*it has a stop sign on each of its four corners*—works the same way but is more complicated because it requires abductive reasoning involving the meaning of *each*, the knowledge for which is stored in the lexical sense for the construction:

Subj has [DirectObj NP [PP Prep [NP each of (Det) Num Noun]]

Interpreting the meaning of this construction requires abductive reasoning—specifically, an understanding that "a stop sign on each of its four corners" means that there are four stop signs, one on each corner.

To recap the main points about the processing of this example:

- Particular linguistic constructions suggest that particular kinds of set-oriented reasoning are needed.
- The meaning procedures encoded in the lexical senses for those constructions trigger the code that carries out that reasoning.
- The reasoning is undertaken as a part of normal TMR creation, during the fourth of the agent's six stages of language understanding: Lexically Triggered Procedural Semantics.
- This reasoning is essential to computing the meaning of the input.
- The result of this reasoning is available in the TMR that seeds learning.
- When the agent reasons about learning ontology, a function in the "Learn property values" script incorporates this knowledge into the ontological concept FOUR-WAY-STOP.

Human: A stop sign is an octagonal red sign with white letters. Interpreting this sentence uses the lexical sense for the construction *a Noun is a Noun*, which maps to the IS-A relation and indicates a concept's ontological parent. The modifiers *red* and *with white letters* are analyzed in the normal way, resulting in the TMR:

```
STOP-SIGN-1
    IS-A                SIGN-1
SIGN-1
    SHAPE               octagonal
    COLOR               red
    HAS-OBJECT-AS-PART  LETTER-OF-ALPHABET-1
LETTER-OF-ALPHABET-1
    CARDINALITY         >1              ; a shorthand for full set notation
    COLOR               white
```

What is noteworthy about this TMR is that it does not directly attach the property values to STOP-SIGN; it attaches them to SIGN. As part of ontology learning—namely, as part of the reasoning carried out by the "Learn property values" script—the agent reinterprets this TMR as the following ontological frame:

STOP-SIGN
    IS-A   SIGN
    SHAPE   OCTAGONAL
    COLOR   red
    HAS-OBJECT-AS-PART   LETTER-OF-ALPHABET-1
LETTER-OF-ALPHABET-1
    CARDINALITY   >1
    COLOR   white

This reasoning uses a straightforward rule that anticipates language inputs of the form "CONCEPT-1 IS-A CONCEPT-2$_{+\text{PROPERTY-VALUES}}$" and transforms it into "CONCEPT-1$_{+\text{PROPERTY-VALUES}}$ IS-A CONCEPT-2." It is the latter knowledge that the agent adds to its ontology.

Human: It is usually about 30 inches across. But it can also be 35 inches wide. The first thing to note about these sentences is that they include details that most people neither know nor would think to mention. However, to optimize an agent's learning in a high-stakes domains like driving, human teachers would be well advised to be as specific as possible. In other domains, relevant scalar values will be common knowledge, making the type of reasoning illustrated by this example widely applicable.

The first and second sentences require different kinds of reasoning that are carried out at different points in the learning process. The first one requires interpreting the meaning of *about* when it modifies a scalar attribute.[14] A meaning procedure attached to the relevant lexical sense of *about* translates the named value into a range surrounding it. The default rule, which is used for objects that are not provided with more specific rules, involves generating a range using 6 percent of the listed value. Here, the result is the range 28.2–31.8 inches. So, after the first sentence, the agent has the following knowledge to seed learning:

STOP-SIGN
    WIDTH   28.2–31.8 (MEASURED-IN INCH)

The next sentence indicates that the width can be 35 inches, resulting in the following knowledge:

STOP-SIGN
    WIDTH   35 (MEASURED-IN INCH)

The agent's "Learn property values" script includes a function to deal with such discrepancies. It learns the WIDTH of a STOP-SIGN as a tuple listing the evidence count, minimum and maximum known values, and the mean: {2, 28.2, 35, 31.6}.

Human: After you stop at a four-way stop, you must decide when to start driving. This sentence presents an ordered pair of events that includes a necessary precondition. Language understanding presents no special challenges and results in the following TMR:

STOP-DRIVING-EVENT-1		; stopping driving
LOCATION	FOUR-WAY-STOP-1	; at a four-way stop
PRECONDITION-OF	MODALITY-1	; is a precondition of
MODALITY-1		; having to decide
TYPE	OBLIGATIVE	
VALUE	1	
SCOPE	DECIDE-1	
DECIDE-1		
AGENT	LEIA-1	
THEME	ASPECT-1.TIME	; when to start driving
ASPECT-1		
PHASE	begin	
SCOPE	DRIVE-1	

When the agent evaluates this TMR using the "Learn property values" script, it reinterprets the PRECONDITION-OF property as conveying the next filler of the SUBEVENTS property of EVENT-AT-A-FOUR-WAY-STOP script. As earlier, this is a simple mapping rule that recognizes a particular shape of a TMR and transforms it into something more suitable for ontology learning.

Human: If there are no vehicles at the intersection, you can go. This input is straightforward in terms of language understanding, resulting in the following TMR:

MODALITY-1		
TYPE	PERMISSIVE	
VALUE	1	; you can
SCOPE	DRIVE-1	; drive
PRECONDITION	LOCATION-1	; if
LOCATION-1		; at the intersection
DOMAIN	INTERSECTION-1	
RANGE	SET-1	; there are no vehicles
SET-1		
MEMBER-TYPE	VEHICLE	
CARDINALITY	0	

When the LEIA analyzes this TMR using the "Learn property values" script, it recognizes a spatially oriented property that is expressed in qualitative rather than quantitative terms: namely, *at the intersection* maps to LOCATION. A standing goal in "Learn property values" is to be on the lookout for such situations since spatial reasoning can require grounding in concrete values of the physical world. When the goal of concretizing a qualitative representation is instantiated, the agent, as always, has the option to wait and see or to ask a clarification question. And, as always, its choice is based on features of the application, such as how proactive a given user wants the agent to be and how recently the agent last asked a question. In this case, the agent takes the initiative and asks a question.

LEIA: How do I tell if a vehicle is at the intersection?   The construction *How do I tell if Clause?* is one of the available realizations of REQUEST-INFO in the LEIA's lexicon. So, the agent can generate this question in the normal way enabled by its language generation engine.

Human: A vehicle is at the intersection if it is adjacent to a stop sign and not moving.   This input provides more information about the vehicle—that it is not moving. It also provides a different qualitative property involving its location: BESIDE, which is the analysis of *adjacent to*. Although BESIDE is more specific than LOCATION, it is still qualitative and, therefore, triggers a new instance of the standing goal of concretizing a qualitative spatial relation. The agent has the same choice as before: to wait and see if the human provides quantitative information or to ask about it. It chooses to ask about it. But it does not ask a general question, it seeks corroboration of the results of its reasoning.

LEIA: I think a vehicle is beside a stop sign when it is no more than 50 inches away. Is that right?   The agent generates this clarification question based on its hypothesis about what *beside* means. It arrives at this hypothesis using a function in "Learn property values." This function consults the agent's ontology to see if the ontological description of a qualitative property suggests how to calculate a quantitative correlate. In this case, the description of BESIDE does have such a function:

BESIDE(A,B) is true iff DISTANCE-SPATIAL(A,B) < 0.25 MAX(A.WIDTH B.WIDTH).

Since the value ranges for the WIDTH property of the concepts STOP-SIGN and ROAD-VEHICLE are available, the LEIA can use the above formula to come up with a number. Since this is a dialog application, the LEIA can check whether the human agrees with its conclusion, which it does by generating the sentence *I think a vehicle is beside a stop sign when it is no more than 50 inches away.*

Human: Yes. When the agent's hypothesis is confirmed, a function in "Learn ontological properties" records the candidate knowledge and "Select ontology learning move" decides to commit it to memory.

Human: If other vehicles arrived at the intersection before you, they have the right of way. This input is needed to round out the scenario but it does not involve any phenomena that we have not already illustrated: the language understanding is straightforward, no special reasoning rules are triggered, conditional statements are a regular part of ontological scripts, and the agent has no need to ask follow-up questions.

At the conclusion of this learning sequence, the LEIA still does not have complete knowledge about how to behave at a four way stop sign: for example, it does not yet know what to do in emergencies, such as in the face of reckless driving. This points to the fact that learning is one thing and assessing whether the agent has sufficient knowledge to function reliably in one or another application is another.

The main takeaway from this extended example is that all of the language processing and learning-oriented demands are covered by procedures in the processing flow shown in figure 7.1. Reasoning rules are recorded as generically as possible and are available for applications of any type and in any domain.

7.2 Part 2: Eventualities in Learning

As the examples above demonstrate, while learning, agents can encounter different kinds and combinations of new information that they understand to varying degrees. The space of learning-oriented eventualities is large. The sections to follow present a starting inventory of eventualities involved in learning lexicon and ontology and the ways that LEIAs handle each one.

7.2.1 Lexicon Learning during Natural Language Understanding

A LEIA reasons about learning new words and phrases at multiple points of the learning process. This section describes the five aspects of lexicon learning that are distributed across the six stages of Natural Language Understanding:

1. Basic Syntax
2. OntoSyntax

3. Basic Semantics

4. Lexically Triggered Procedural Semantics

5. Extended Semantics

6. Situational Reasoning

Posit Senses for Unknown Words   During OntoSyntax, the agent posits senses for unknown words and for known words that are used in a syntactically unexpected way. As an example of the latter, if the input is *Casey is walking her dog* but the lexicon contains only an intransitive sense of *walk*, then a transitive sense needs to be learned. During Onto-Syntax, semantics is not yet being considered, so if the lexicon contains a sense that is syntactically appropriate but semantically wrong, that will not yet be detected; it will be detected and treated during Extended Semantics using the function "Resolve incongruities using a new lexical sense."

When the agent detects an unknown word or word usage, it first determines if it knows any morphologically related words that might be semantically related. For example, if it detects the unknown verb *fuel* and knows the noun *fuel*, might the two be related? Hypothesizing that they might be is likely what people do, so it is a natural first step for the agent as well. Of course, there is no guarantee that morphologically related words will be semantically related because of complications arising from polysemy, homography, semantic shift, and the use of words in idiomatic expressions. This points to the fact that we cannot make natural languages any simpler or more orderly than they are; we can only equip agents to reason as best they can given this state of affairs.

The following subsections explain how agents posit lexical senses for unknown nouns, adjectives, and verbs during OntoSyntax.

Example Presentation Note. This section illustrates the kinds of lexical senses that the agent can posit by showing how they will ultimately be used in text meaning representations (TMRs). Procedurally, this is jumping the gun because positing new senses occurs during OntoSyntax, which is the stage of language understanding before TMRs are created. However, this foreshadowing is justified because it will concretize what would otherwise be an abstract, technically dense presentation.

Table 7.3
Examples of verbal senses being used to analyze unknown nouns.

Verb type	Verbal example	Morphological analysis	Nominal example
intransitive	People snooze.	no morphology	snooze
	People sunbathe.	-ing	sunbathing (by people)
	Connie sunbathes.	-er (agentive)	sunbather
transitive	Workmen construct buildings.	-ion	construction (of buildings) (by workmen)
	Ed mows lawns.	-er (agentive)	mower (of lawns)
	Machines mow lawns.	-er (instrumental)	mower
ditransitive	People donate money to charity.	-ion	donation (of money) (to charity) (by people)

Posit Senses for Unknown Nouns   Unknown nouns may or may not be related to known word senses. If a noun is related to another word, it is usually a verb, and the base form of the noun might look the same as the base form of the verb (walk$_{Verb}$, walk$_{Noun}$) or be derivable using a rule of derivational morphology (sunbathe$_{Verb}$, sunbather$_{Noun}$). If the agent knows a verb that might be related to the unknown noun it is processing, the verb's dependencies affect what the candidate noun sense will look like, as illustrated in table 7.3.

As an example of learning a noun sense from a verb sense, assume that lexicon contains the verb *construct* but not the noun *construction*, which the agent encounters in the noun phrase *the construction of the building*. The results of morphological analysis feed into a mapping function that converts the verb sense construct-v1 into the noun sense construction-n1 as follows.

```
construct-v1
   definition      to build (a physical action)
   example         The workers are constructing a new hospital.
   syntax-type     v-trans
   output-syntax   CL
   syn-struc
      subject      $var1
      v            $var0
      directobject $var2
```

```
    sem-struc
      BUILD
         AGENT          ^$var1
         THEME          ^$var2
construction-n1
    definition          -
    example             the construction of the building  ; the input triggering the learning
    comments            auto-learned sense from construct-v1
    syntax-type         n+opt-pp+opt-pp
    output-syntax       N
    syn-struc
      n                 $var0
      pp (opt+)
         prep           $var1 (root 'of')
         np             $var2
      pp (opt+)
         prep           $var3 (root 'by')
         np             $var4
    sem-struc
      BUILD-1
         AGENT          ^$var4
         THEME          ^$var2
      ^$var1,3 null-sem+
    meaning-procedures
      vet-learning  BUILD-1
```

For the input *the construction of the building*, the newly learned sense construction-n1 will allow the agent to produce the TMR

```
BUILD-1
    THEME      BUILDING-STRUCTURE-1
    HAS-MP     vet-learning (BUILD-1)
```

Like all candidate senses that are automatically learned, this one is appended with a meaning procedure that indicates that this hypothesized analysis needs to be semantically vetted, which will be done later on, during Lexically Triggered Procedural Semantics. Reasons why the analysis might not be correct include:

- The noun might not be associated with any verb sense at all.

- The noun might be associated with a verb sense but not in a way predicted by the rule that fired: for example, a *cooker* is an appliance, not a person who cooks.

- The noun might be associated with a verb sense but one that is missing from the agent's lexicon.

At this early stage of language analysis, the agent is preparing for all reasonable eventualities in order to avoid backtracking. Therefore, an additional candidate sense of the unknown noun is posited that maps to a generic EVENT.[15] This sense is available if the more specific candidate mapping turns out to be semantically incompatible with the overall context.

```
construction-n2
    definition      -
    example         the construction of the building
    syntax-type     n-bare
    comments        auto-learned from scratch
    output-syntax   N
    syn-struc
       n            $var0
    sem-struc
       EVENT-1
    meaning-procedures
       seek-specification EVENT-1
```

This sense is appended with a meaning procedure called seek-specification that indicates that the semantic mapping needs to be further specified if the underspecified interpretation is not actionable or if the agent wants to permanently store the newly learned word in its lexicon (it will not permanently learn such an underspecified meaning).

If, for an unknown noun, there are no related meanings in the lexicon, two candidate senses are posited—one mapping to OBJECT and the other to EVENT.

Posit Senses for Unknown Adjectives Syntactically, adjectives modify nouns. Semantically, they typically map to a PROPERTY, and the noun they modify fills the DOMAIN slot of that property. The RANGE depends on the adjective's meaning. For example, the meaning of *tall* is "HEIGHT .8," so a *tall tree* is

```
TREE-1
    HEIGHT  .8
```

If the agent encounters the noun phrase *a quick-witted systems engineer* but does not know the adjective *quick-witted*, it will posit the following lexical sense:

```
quick-witted-adj1
    definition      -
    example         a quick-witted systems engineer
```

```
  comments         auto-learned from scratch
  syntax-type      plain-adj
  output-syntax    N
  syn-struc
    mod    $var0
    n      $var1
  sem-struc
    PROPERTY-1
      DOMAIN      ^$var1
  meaning-procedures
    seek-specification PROPERTY-1
    seek-specification PROPERTY-1.RANGE
```

For a *quick-witted systems engineer*, this lexical sense will allow the agent to generate the TMR

```
PROPERTY-1
  DOMAIN    SYSTEMS-ENGINEER-1
  HAS-MP    seek-specification (PROPERTY-1)
  HAS-MP    seek-specification (PROPERTY-1.RANGE)
```

The new adjective sense includes two meaning procedures, whose calls are carried into the TMR: one for determining which property the adjective maps to and the other for determining its value. A noteworthy aspect of learning properties is that more examples will not help—the agent needs a definition to learn a property's meaning, which can be provided by a human or a text.

Posit Senses for Unknown Verbs   Unknown verb senses involve more even-tualities. A verb sense could need to be learned either because the lexicon lacks any sense of the verb at all or because the lexicon lacks a sense that aligns with the syntactic structure of the input. (Remember, the agent is not yet evaluating whether the semantics of known senses fit the context.) The agent is supplied with templates to support the learning of typical verb types. For example, below is the template for learning a transitive verb sense.

```
[new-transitive-verb]-v1
  definition       -
  example          [the sentence that triggers the learning]
  comments         a transitive verb being learned on the fly
  syntax-type      v-trans
  output-syntax    CL
```

```
syn-struc
    subject        $var1
    v              $var0
    directobject   $var2
sem-struc
    EVENT-1
        CASE-ROLE-1    ^$var1
        CASE-ROLE-2    ^$var2
meaning-procedures
    seek-specification EVENT-1
    seek-specification CASE-ROLE-1
    seek-specification CASE-ROLE-2
```

The verb maps to a generic EVENT whose case roles are generic CASE-ROLES. All of these are appended with meaning procedures that indicate what needs to be further specified if the underspecified analysis does not render the TMR actionable or if the agent wants to store the learned sense to its lexicon.

As an example, if the agent receives the input *The magician gobsmacked the audience,* but it does not know the verb *gobsmack,* it will use the new-transitive-verb template to generate the following TMR.

Example Presentation Note. The TMRs in this section are abbreviated—stripped of information about tense, aspect, and so on—in order to focus on what is most important for the current discussion.

```
EVENT-1
    CASE-ROLE-1    MAGICIAN-1
    CASE-ROLE-2    AUDIENCE-1
    HAS-MP         seek-specification (EVENT-1)
    HAS-MP         seek-specification (EVENT-1.CASE-ROLE-1)
    HAS-MP         seek-specification (EVENT-1.CASE-ROLE-2)
```

The reason why specific case roles are not assigned by default is that there are no reliable defaults prior to semantic analysis. For example, whereas the magician in our example fills the AGENT slot, if the example were *The magic show gobsmacked the audience,* the magic show would fill the INSTRUMENT slot.

The above treatment applies to wholly unknown verbs. However, some unknown verbs are potentially related to a known verb or noun, which offers an inroad to a better analysis. Following Goldberg (2019), we refer to such analyses as *coercion*—that is, they involve coercing a known meaning to accommodate an unknown usage. Some details about coercion:

- Each type of coercion is implemented using rules that map known senses into new ones.

- When the meaning of an unknown word is coerced from a known meaning, the TMR carries a call to a meaning procedure to semantically vet the analysis.

- Since coercion might misfire, whenever a coerced analysis is posited, at least one generic analysis is also added to the candidate space. The semantic suitability of the coerced analysis will be assessed as part of semantic analysis.

Below we illustrate seven types of coercion using examples. Rather than providing the source and target lexical senses for each example, we explain the generalization in plain English and illustrate it with a sample sentence and its TMR. Only the correct TMR is shown for each case even though, in a LEIA's actual operation, it generates candidate interpretations using all candidate senses and then prunes out the incorrect ones.

Type 1: The unknown verb is ditransitive, and ditransitive verbs, by default, indicate TRANSFER-POSSESSION. Psycholinguistic research has shown that the default interpretation for ditransitive verbs is TRANSFER-POSSESSION with the case roles AGENT, THEME, and BENEFICIARY (Goldberg, 2019). So, given the sentence *She lobbed the ball to Harry*, and assuming that the agent does not know the verb *lob*, it will select the ditransitive new-word-learning template and generate the following TMR from our example:

TRANSFER-POSSESSION-1
    AGENT   HUMAN-1
    THEME   BALL-1
    BENEFICIARY   HUMAN-2
    HAS-MP   vet-analysis (TRANSFER-POSSESSION-1)
HUMAN-1
    HAS-GENDER   female
HUMAN-2
    HAS-NAME   "Harry"

Note that the concept TRANSFER-POSSESSION is defined broadly enough to make it semantically correct, albeit underspecified, in this context.

Type 2: The verb is known but with different syntactic expectations. For
example, the lexicon might contain the verb *think* as used in *He thought about
it* but not as used in *He thought to come*. Similarly, the lexicon might contain
the verb *volunteer* as used in *He volunteered to go* but not as used in *She volunteered him to go*. In such cases, the agent (a) posits a lexical sense with the
syntactic structure attested in the parse of the input, (b) uses the same event
mapping as for the known verb sense, and (c) changes the specific case roles
of the known sense into generic CASE-ROLES for the learned one. For *She volunteered him to go*, the TMR generated from the coerced sense is:

VOLUNTEER-1
    CASE-ROLE-1   HUMAN-1
    CASE-ROLE-2   HUMAN-2
    CASE-ROLE-3   MOTION-EVENT-1
    HAS-MP   vet-analysis (VOLUNTEER-1)
    HAS-MP   seek-specification (CASE-ROLE-1)
    HAS-MP   seek-specification (CASE-ROLE-2)
    HAS-MP   seek-specification (CASE-ROLE-3)
HUMAN-1
    HAS-GENDER   female
HUMAN-2
    HAS-GENDER   male

Although this TMR is underspecified in multiple ways, consider what it *does*
convey: there is an instance of volunteering, the two humans in question
are involved, and the volunteering involves a motion event. In terms of the
agent's goal of arriving at actionable interpretations of inputs, this is a lot
of useful information.

It is important to understand why the case roles are underspecified. It is
not possible to predict whether the syntactic construction *Subj V DirectObj
XComp*, when used with an unknown main verb, will be subject-controlled
or object-controlled. This construction is subject-controlled in *She promised
him to go*, which involves her going, whereas it is object-controlled in *She
told him to go*, which involves him going. *She volunteered him to go* happens
to be object-controlled.

The ontological concept VOLUNTEER is defined such that the AGENT is the
individual that will carry out the volunteered-for activity, and the THEME is
the activity itself. For the input *She volunteered him to go*, *she* is *not* the AGENT
of the VOLUNTEER event; *she* fills the CAUSED-BY slot. So, the ideal TMR, which
requires more information to generate, is:

VOLUNTEER-1
  CAUSED-BY   HUMAN-1
  THEME   MOTION-EVENT-1
MOTION-EVENT-1
  AGENT   HUMAN-2
HUMAN-1
  HAS-GENDER female
HUMAN-2
  HAS-GENDER male

Whereas unknown word usages like those above reflect basic knowledge that the LEIA should have, others reflect so-called *linguistic creativity*—usages that we would not expect to be stored in a person's or agent's knowledge base. People play with language all the time to create novel, snappy, and sometimes humorous utterances.[16] Since this playing tends to follow predictable patterns—and, therefore, is actually less creative than the term implies—LEIAs can be prepared to interpret many such cases, including those listed as types 3–7 below, which were inspired by Goldberg (2019). When, by contrast, the agent cannot make a better-than-baseline hypothesis about the semantics of a new word sense, it uses the generic OBJECT or EVENT meaning that is the default for new-sense learning.

In reading the following examples, remember that creative language use is atypical, so the examples do not sound like everyday English; but they could absolutely be uttered by native speakers.

Type 3:   A known noun is used as a transitive verb, as in *Louise is forking the pie crust*. In this case, the agent instantiates an EVENT for which:

- The meaning of the subject fills the AGENT slot since it is animate; if it were inanimate, it would fill a generic CASE-ROLE slot.
- The meaning of the direct object fills the THEME slot.
- The meaning of the known noun fills the RELATION slot; for this example, INSTRUMENT would be more precise, but it is not a reliable guess in the general case.

The resulting TMR for *Louise is forking the pie crust* is as follows. It says that Louise is doing something to a pie crust involving a fork.

EVENT-1
  AGENT   HUMAN-1
  THEME   PIE-CRUST-1
  RELATION FORK-1

HAS-MP	seek-specification (EVENT-1)
HAS-MP	seek-specification (EVENT-1.RELATION)

HUMAN-1

HAS-NAME	"Louise"

Type 4:   A known noun is used as an intransitive verb, as in *Hank is driveway-ing again.* The TMR says that Hank is doing something related to a driveway.

EVENT-1

AGENT	HUMAN-1
RELATION	DRIVEWAY-1
HAS-MP	seek-specification (EVENT-1)
HAS-MP	seek-specification (EVENT-1.RELATION)

HUMAN-1

HAS-NAME	"Hank"

This could be said, for example, of a person who is known to sweep his driveway a lot.

Type 5:   A known noun is used as a ditransitive verb, as in *Mary rulered Tim a piece of chalk* or *Mary rulered a piece of chalk to Tim*, both of which result in the same analysis.[17]

TRANSFER-POSSESSION-1

AGENT	HUMAN-1
THEME	CHALK-FOR-BLACKBOARD-1
INSTRUMENT	LENGTH-MEASURING-ARTIFACT-1
BENEFICIARY	HUMAN-2
HAS-MP	vet-analysis (TRANSFER-POSSESSION-1)

HUMAN-1

HAS-NAME	"Mary"

HUMAN-2

HAS-NAME	"Tim"

Type 6:   A known intransitive verb indicating a physical action is used in a caused-motion construction, as in *Mary sneezed a piece of chalk to Tim.*[18] The TMR says that a piece of chalk moved, it was caused by Mary's sneeze, and the beneficiary of the chalk's motion was Tim.

MOTION-EVENT-1

THEME	CHALK-FOR-BLACKBOARD-1
BENEFICIARY	HUMAN-2
CAUSED-BY	SNEEZE-1
HAS-MP	vet-analysis (MOTION-EVENT-1)

SNEEZE-1
   EXPERIENCER   HUMAN-1
HUMAN-1
   HAS-NAME   "Mary"
HUMAN-2
   HAS-NAME   "Tim"

For the input *Mary sneezed a piece of chalk onto the floor*, FLOOR would be the DESTINATION of the MOTION-EVENT.

Type 7:   A known transitive verb is used in a resultative construction, as in *Mary kissed Tim conscious*.[19] The TMR says that Mary kissed Tim, before the kiss (the PRECONDITION) Tim was unconscious, and after the kiss (the EFFECT) he was conscious.

KISS-1
   AGENT   HUMAN-1
   BENEFICIARY   HUMAN-2
   PRECONDITION   STATE-OF-CONSCIOUSNESS-1
   EFFECT   STATE-OF-CONSCIOUSNESS-2
   HAS-MP   vet-analysis (KISS-1)
HUMAN-1
   HAS-NAME   "Mary"
HUMAN-2
   HAS-NAME   "Tim"
STATE-OF-CONSCIOUSNESS-1
   DOMAIN   HUMAN-2
   RANGE   unconscious
STATE-OF-CONSCIOUSNESS-2
   DOMAIN   HUMAN-2
   RANGE   conscious

To generalize, in some cases of coercion, the agent arrives at an analysis that is similar to what a person would come up with, as when rulering a piece of chalk to someone. In other cases, the agent arrives at a vaguer interpretation because specific reasoning about the world is needed for a more precise analysis. For example, *drivewaying* is limited to typical actions that could be performed on a driveway, such as sweeping it, hosing it, or resealing it. Whether the agent pursues a more specific interpretation depends on its priorities in a given application.

Since positing lexical senses via coercion occurs at the very start of language analysis, transformations—passivization, creating imperatives and questions, and so on—can apply to the newly posited senses in the normal way.

Hypothesize a Multiword Analysis   Language is full of multiword expressions, and any unknown word that is encountered might be part of one. For example:

- One or more nouns in a nominal compound can be unknown: *creature comforts, golf cart.*
- One or more words in an Adjective Noun collocation (which can include any number of adjectives and nouns) can be unknown: *differential reinforcement, differential reinforcement schedule.*
- One or more of the nouns in a noun phrase that includes a prepositional phrase might be unknown: *eye of the needle.*
- An unknown verb can be followed by a syntactically ambiguous sequence of constituents. For example, *They bricked up the building* contains the phrasal verb *bricked up* and must be interpreted as "Subj V Particle Direct-Obj." It would be incorrect to analyze this sentence as "Subj V PP" and to attempt to learn the verb *to brick* without the particle.
- Idiomatic sentences, sayings, and proverbs can contain unknown words: *Absence makes the heart grow fonder.*

Moreover, combinations of known words can form multiword expressions that are not yet recorded in the lexicon. For example, the verbs *have, take,* and *make* are used in many idiomatic expressions each: for example, *take a shower<look, break, message, whiff, taxi>.* It can be tricky to automatically detect whether an unknown word is free-standing or is part of a multiword expression.

The function "Hypothesize a multiword analysis" uses a set of heuristics to determine whether a multiword analysis is plausible. If it is, the agent treats it as the preferred analysis. There is, however, one downside to preferring multiword analyses, which involves scoring during lexical disambiguation. Our scoring system prefers multiword analyses over compositional ones. So, if, while being taught about physiology, the agent learns that the multiword expression *body part* means BODY-PART—which refers to an animal's body part—then this analysis will be favored over compositional analyses of *body part* during lexical disambiguation. The problem is that *body part* can also refer to a part of a car. What to do? The most straightforward solution is to batch-vet lists of learned multiword expressions with an eye toward spotting ambiguity. If a learned multiword expression is found to be ambiguous, then lexical senses for its other meaning(s) should be acquired as well—in our example, VEHICLE-PART. Then, each time the agent encounters *body part*, both multiword analyses will get the same multiword scoring bonus, and contextual reasoning will be needed to select between them.

Preparing agents to identify and learn multiword expressions requires judgment and restraint. It would be impractical to assume that every syntactic constituent at every level of every parse might be a multiword expression. Currently the agent hypothesizes a multiword expression only if one of the following conditions is met:

1. A noun phrase of any shape is used in an utterance that asserts that it is the topic of teaching, such as *Today I'll be teaching you about NP*. This situation is recognized when the TMR contains a TEACH event whose THEME is filled by the meaning of a noun phrase that contains the unknown word. This heuristic is particularly useful for entities whose names include a postmodifier since it asserts that the postmodifier is part of the entity's name: *Today I'll be teaching you about systemic sclerosis with esophageal involvement*.

2. A noun phrase of any shape is the subject of a copular construction of the form *Subj is a(n) NP*: *Systemic sclerosis with esophageal involvement is an idiopathic disease*.

3. A noun phrase that includes an unknown noun is used as a nominal complement: *A particularly interesting disease is systemic sclerosis with esophageal involvement*.

The agent uses rules like these to hypothesize multiword expressions in any application. In corpus-based learning specifically, the agent can investigate which word combinations are likely to be multiword expressions based on their frequency, their inclusion in lists of terminology, and so on.

Leverage Selectional Constraints in the Input The selectional constraints of known concepts can sometimes help to narrow down the interpretation of unknown words. The best case is when:

- The unknown word is a plain noun like *eggnog*.
- It is the only unknown word in the sentence or within an independent clause in the sentence.
- The unknown word serves as a dependent of a known verb that is unambiguous in the given syntactic environment: *Joyce was drinking eggnog all evening*.
- The selectional constraints on the argument filled by the unknown word are narrow: the THEME of DRINK must be BEVERAGE.

Given this information, the agent can confidently guess that *eggnog* is some sort of BEVERAGE. Of course, that is not a maximally precise description, but it is a useful first approximation that can seed further property-based learning.

In less optimal contexts, like *Joyce loves eggnog*, it can be impossible to glean useful semantic information from a single example—after all, one can love anything. When the agent reasons about what to do next during lexicon learning, it recognizes this situation and can launch data-driven analysis to try to collect examples that include verbs with narrower selectional constraints.

Disambiguate Using Ontology Learning Preferences   When agents are learning ontology from language inputs, the fact that they are aware that they are learning ontology helps them to make certain kinds of lexical disambiguation decisions. Four ontology-learning generalizations are as follows:

Prefer the generic over the specific interpretation of NPs with *a* and *the* in the absence of counterevidence.   For example, a learning session about clinical medicine might include the statements *Systemic sclerosis can affect the esophagus* and *A Heller myotomy is a surgical treatment for achalasia*. Here, *the esophagus* and *a Heller myotomy* are generic. An example of counterevidence that blocks the generic reading is when the agent's coreference resolution procedures identify a specific sponsor for the NP. For example, a published case study might include multiple references to a specific patient's esophagus.

Prefer the generic over the specific interpretation of *you* in the absence of counterevidence.[20]   For example, if a teaching clinician says, *When you interact with a patient, you should . . .* , this is usually generic advice that applies to any physician. An example of counterevidence is if a teacher juxtaposes generic advice with advice catered to a specific trainee.

Prefer domain-specific interpretations of lexically ambiguous words when engaging in domain-specific, deliberate learning.   When an agent is deliberately learning about a particular domain, that means that it knows which segments of ontology are relevant. Therefore, given ambiguous words, it can prefer the domain-relevant interpretation. For example, in the context of clinical medicine for humans, *body* refers to a live human body, not the body of an animal, a corpse, or a part of a vehicle. Outside of learning contexts, automatically detecting the domain of an utterance can be difficult because people switch topics all the time.[21]

Prefer script-relevant interpretations of words that are ambiguous or unknown. We saw the need for this in the example of a patient learning about achalasia (section 7.1.2). The input was *You have achalasia,* in which *have* is ambiguous and *achalasia* is unknown. From this, the agent figured out that achalasia is a disease that it is suffering from. But how did it do that?

This dialog takes place at a doctor's office, which means that the DOCTORS-OFFICE script is instantiated. All of its subevents are, therefore, salient. One of those is DIAGNOSE-DISEASE, which is what is happening in our example. The agent needs to make the connection between the language input *You have achalasia$_{unknown-word}$* and the event DIAGNOSE-DISEASE. It does this by working through all of the lexical senses for *have* that syntactically match the input, semantically match the input with respect known case-role fillers, and meet all of the requirements computed using meaning procedures. One candidate analysis of *have* that meets all of these requirements is *Subj$_{ANIMAL}$ has DirectObj$_{DISEASE}$*: It is transitive, *you* fulfills the ANIMAL constraint on the subject, and the speaker is a doctor. So, the agent asks itself, "If the meaning of the unknown word aligned with the semantic constraint of the slot it is filling—here, if achalasia were a DISEASE—then would the meaning of this construction match any of the events in the script I am participating in?" In this case, the answer is yes because the semantic interpretation of the construction *Subj$_{ANIMAL}$ have DirectObj$_{DISEASE}$* is DIAGNOSE-DISEASE, whose THEME is the DISEASE that is EXPERIENCED-BY the given ANIMAL. This precisely matches the corresponding event in the DOCTORS-OFFICE script, thus signaling to the agent that this is probably the intended meaning. Based on this hypothesis, it learns the mapping of the word *achalasia* to a new concept called ACHALA-SIA, which is a child of DISEASE.

All of the abovementioned disambiguation functions involve preferring certain candidate meaning representations over others based on the agent's understanding of various aspects of the context.

Resolve Incongruities via New Sense Learning Midway through language understanding—specifically, at stage 5, Extended Semantics—the agent may encounter the eventuality that all candidate TMRs suffer from semantic incongruity. This means that none of them completely fulfills the ontological expectations that events impose on their case-role fillers. Given the many potential sources of incongruity—humor, misspeaking, metaphorical extension, and so on—LEIAs need heuristics to decide which explanation is most likely so that they can launch the associated repair action. When they are engaged in ontology learning, if an incongruity can be resolved by positing a new word sense, that repair action is preferred because polysemy is rampant in language. No matter how many senses of a word an agent knows, it can always encounter a new one. (Here we are talking about unknown senses that are syntactically identical to known ones—that is, they are of the same part of speech and exhibit the same syntactic behavior. We are not talking about coercing known senses into new syntactic uses.)

Although the always-looming possibility of a new sense might seem to suggest that the agent should actively anticipate that every word in every text might reflect a new sense, that would be impractical: it would not only explode the number of candidate analyses of any input, it would also always result in an analysis that contained exclusively new senses, which is hardly likely to be the right answer. To make the agent behave sensibly, we require it to have a good reason to posit a new sense of a known word. These reasons are modeled as heuristics: if a TMR with a semantic incongruity has any of a list of particular features, then resolve the incongruity by positing a new word sense in a particular way. We will illustrate this using two examples.

A common type of incongruity occurs when a known verb sense is syntactically appropriate but requires an agentive subject, whereas the input's subject is not agentive. For example, the lexical sense for *advise* that covers the input *Their manager advises employees not to be late* will not cover *This booklet advises employees not to be late* because booklets are not animate so they cannot be agents. When a LEIA identifies such a case, it posits a new word sense that is the same as the known sense except that the AGENT case role is changed to INSTRUMENT. This new sense explicitly disallows animate subjects.

Another example in which lexicon learning can help to resolve an incongruity occurs when a known word is defined in a way that conflicts with the agent's understanding of it. For example, a teacher of linguistics might introduce the concept of syntactic trees by saying, *A tree is a visual representation of the syntactic structure of a sentence*. The TMR of this sentence will include

TREE-1
    IS-A   REPRESENTATIONAL-OBJECT-1

This description clashes with the LEIA's ontological knowledge, which says that the ancestry of TREE is TREE < PLANT < ANIMATE < PHYSICAL-OBJECT. The agent knows to look out for such situations because one of the functions in "Resolve incongruities" directs it to do so as part of expectation-oriented cognitive modeling.

7.2.2 Learning Ontology and Residual Aspects of Lexicon

Ontology learning can be triggered in various ways: by language input; by input from other channels of perception, such as vision and interoception; or without any external input, since the LEIA has a standing goal of ontology revision as an aspect of memory management. When a new concept is learned as part of processing an unknown word or expression, the lexical sense for that word or expression must use the newly created concept. In such cases, lexicon learning and concept learning go hand in hand.

Create and Anchor a New Concept   The agent does not create a new concept for every new thing it encounters for two reasons. First, many encountered things are best understood as instances of existing concepts. For example, the word *settee* is close enough to a standard *couch* to map to the concept COUCH. Second, the agent cannot learn about every novel stimulus at the same time. In purely task-oriented applications, it needs to focus on its task and ignore extraneous stimuli; and in learning-inclusive applications, it needs to learn entities in an orderly manner. So, the first decision in the function "Create and anchor a new concept" is whether or not to actually create a new concept.

The agent does *not* create a new concept if:

> A newly learned word is described as having a known meaning: for example, *Stunning means beautiful.* In this case, the lexical sense for the new word *stunning* uses the same concept mapping as the known word *beautiful.*

> *or*

> A coercion rule that maps a known word sense into a new one retains a precise concept mapping: for example, if a nominal sense of *walk* is learned from the intransitive verbal sense of *walk*, it will be mapped to same concept—WALK.

> *or*

> Because of decision-making specific to the application, the agent decides that it is not worth it to learn a new concept to reflect the meaning of the new word or phrase—a mapping that uses an existing concept is sufficient.

If none of these conditions holds, the agent creates a new concept to accommodate the newly encountered entity.

When the agent creates a new concept, it names it using simple naming conventions that are based on the word or expression itself. For example, when learning the expression *systemic sclerosis*, it creates the concept SYSTEMIC-SCLEROSIS.

A concept thus created might end up being redundant. That is, it might end up overlapping with an existing concept that has a close enough meaning to accommodate the new word or expression. However, the agent cannot know this before it has learned enough about the entity's meaning to recognize the overlap. So, at this point in processing, it goes ahead and posits a new concept and then later on, during "Select ontology learning move," it analyzes whether the candidate concept might be redundant.

All new concepts need to be anchored to the appropriate parent in the ontology. If the given input does not include definitional information, then the concept is tentatively anchored at the highest level: it is made a child of EVENT or OBJECT, as applicable. If, by contrast, the input suggests a more specific parent, then the agent uses that instead. In the best case, the input will convey the ideal parent for the new concept. For example, given the input *A cross clamp is a surgical instrument used to . . .* , the agent will make CROSS-CLAMP a child of SURGICAL-INSTRUMENT. However, people often provide definitions that point to a more distant ancestor, such as *A cross clamp is a tool that . . .* , which would result in CROSS-CLAMP becoming an ontological *child* of TOOL, rather than its *grandchild*. Yet another possibility is that the definition is expressed in vague terms, requiring additional reasoning to identify the concept's parent: for example, *A cross clamp is something surgeons use to . . . (something* maps to OBJECT). The complications of learning an ontological hierarchy from language have been well known for decades (Ide & Véronis, 1993), and there is no simple solution. However, there are practical inroads, such as collecting multiple definitions from text corpora and terminological repositories and then using the most specific hierarchical information available.

Although one might think that it would be useful to seed the ontology using wordnets and terminological "ontologies," which typically include no property-based descriptions, this would be a bad idea for two reasons. First, in order to be most useful for agent reasoning, ontological concepts need to be described by property values that distinguish their meaning from the meaning of their parents and siblings. So, long lists of uninterpreted entities are, at best, only raw material for building an agent's lexicon and ontology. Second, many terminological resources are built with design features that are incompatible with a LEIA's ontology.

As an illustration of the latter, consider the examples discussed in Nirenburg, McShane, and Beale (2009). Two large and well-known resources in the medical domain are MeSH, the National Library of Medicine's hierarchical tree of medical subject headings, and Metathesaurus, the National Library of Medicine's collection of hundreds of thousands of medical terms along with their synonyms and morphological variants. Although Mesh and Metathesaurus contain a wealth of terminology, they were developed for librarians, whose needs are very different from those of agent systems. For example:

- These resources contain English expressions, including extensive nests of synonyms, which introduces aspects of language that the LEIA's language-independent ontology avoids.
- Entities are described by very few properties; and, at the time of our analysis (in 2005), 61 percent of entities had no properties at all.
- There is no division between concepts (abstractions) and instances (real-world objects and events).
- The IS-A relation is loosely defined, so concepts can have many parents that represent quite diverse semantic relationships.
- At the time of our analysis, there were many errors that would need to be cleaned manually. For example, over fourteen thousand concepts were parents of themselves.

In sum, terminological resources are best used as checklists for agents at run-time or as a reference for people during manual knowledge acquisition. LEIAs, by contrast, will build their ontology based on input that is selected by human teachers or by a well-planned progression of data-driven learning tasks.

The special case of learning the meaning of modifiers   Earlier we pointed out that learning the meaning of modifiers, which are initially mapped to the generic PROPERTY, is quite different from learning the meanings of OBJECTs and EVENTs. Properties provide the building blocks for describing OBJECTs and EVENTs and, although there is no perfect inventory, the number should be constrained. The agent should not, for example, create a new property for every new adjective it encounters because many can be described using existing ATTRIBUTEs and RELATIONs. For example, all words and phrases that express aesthetic evaluations are described using AESTHETIC-ATTRIBUTE—*beautiful* (.8), *plain* (.5), *ugly* (.2), *hideous* (0), and so on. So, if the agent encounters the new word *dazzling*, it should not invent a scalar attribute like DAZZLINGNESS to accommodate just this one word; it should describe it as "AESTHETIC-ATTRIBUTE 1."

As a reminder from section 3.2.1, properties make it into the ontology for a variety of reasons.

- Some properties are widely understood to be a core aspect of any world model, such as

  - the case roles: e.g., AGENT, THEME, INSTRUMENT;
  - properties of physical objects: e.g., HEIGHT, WEIGHT, COLOR;
  - properties of physical events: e.g., VELOCITY, DURATION, START-TIME, CAUSED-BY; and
  - complex properties relating to how people think about general-domain entities: e.g., HAS-SPOUSE, CITIZEN-OF, MARITAL-STATUS.

- Other properties reflect the mental models of domain experts and are invented during knowledge engineering. For example, clinicians think about diseases in terms of SUFFICIENT-GROUNDS-TO-DIAGNOSE, SUFFICIENT-GROUNDS-TO-TREAT, HAS-TYPICAL-SYMPTOM, and so on.[22]

So, when, if ever, should the agent independently create a new property? The most reliable answer is *never*—at least not without a human's approval. If an agent is learning about medicine from a human teacher who talks about *viscous blood*, and the agent does not know what *viscous* means, it needs to find out. The process of finding out could be primitive, like asking *What does viscous mean?*, or it could be more sophisticated. A more sophisticated approach would involve consulting various online resources. For this example, the agent could look up *viscous* in online dictionaries and/or thesauri, generate a tentative lexicon entry, and then ask the human if its reasoning is correct. If a definition includes *sticky*, a word in the agent's lexicon that maps to the scalar attribute STICKINESS, then this is a candidate mapping. But since the agent is learning about a specialized domain, it can also hypothesize that this might be a new property; and, if it is, it should be named using the nominal counterpart of viscous, which is also available in online resources—*viscosity*. So, a good response from the agent would be: *Should viscous be described using* STICKINESS *or is* VISCOSITY *a special property?* In this case, we would expect the human to tell the agent to add the new property VISCOSITY.

Another clue that the ontology might need a new property is that the agent encounters a possessive construction or a nominal compound that it is not able to analyze precisely. Possessive constructions and nominal compounds are semantically underspecified. To know which ontological relation is intended, one must know which meanings of the potentially ambiguous words are being juxtaposed and how those meanings most naturally fit together based on world knowledge. Whereas *Alice's mother* indicates OFFSPRING-OF, *Alice's teeth* indicates HAS-OBJECT-AS-PART, and *Alice's toothache* indicates EXPERIENCER-OF. Similarly, whereas *table lamp* indicates BELOW-AND-TOUCHING, *reading lamp* indicates INSTRUMENT, and *copper lamp* indicates MATERIAL-OF.

Many relations that can be expressed by nominal compounds and possessives are accounted for explicitly in two of the agent's knowledge bases.

- Those that combine a fixed word with a concept-constrained variable are recorded in the lexicon: for example, N_{FISH} *fishing* is used to analyze inputs like *salmon fishing* as "FISHING-EVENT (THEME SALMON)."

- Those that combine two concept-constrained variables are recorded in a separate knowledge repository: for example, N_{FOOD} $N_{PREPARED\text{-}FOOD}$ is used to analyze inputs like *spinach lasagna* as "LASAGNA (HAS-OBJECT-AS-PART SPINACH)."[23]

If a nominal compound or possessive construction is explicitly covered by stored knowledge, the agent generates the associated specific analysis and its work is done. By contrast, if an input does not match any stored construction, the agent analyzes the relation using one of two underspecified properties: LINGUISTIC-POSSESSIVE or NOMINAL-COMPOUND-RELATION, depending on the nature of the input. Although, in general, the ontology is no place for things linguistic, making reference to syntax in these concepts is a tactically motivated exception since the nature of the original construction can help to narrow down the intended interpretation. For example, whereas *grandfather's car* is a car possessed by grandfather, *a grandfather car* is a car that is stodgy, not hip.[24]

Returning to ontology learning: if the agent encounters a possessive construction or a nominal compound for which it cannot produce a confident and semantically specific analysis, this is a good reason to ask its human what to do—which can result in using an already-known property or creating a new one.

Yet another situation in which an agent cannot know if a new property is needed is when it learns a new EVENT, since some events result in a state that should be recorded in the ontology. For example, MARRY results in the HAS-SPOUSE relation, and BEAR-OFFSPRING results in the HAS-OFFSPRING relation. Similarly, passing a driver's test, buying a house, and having a mortgage are associated with states that are important enough to be part of the ontology:

"HAS-DRIVERS-LICENSE yes," "IS-HOMEOWNER yes," "HAS-MORTGAGE yes." By contrast, if you thatch your lawn, dust the living room, or spackle the bathroom wall, these do not result in the memorable states of your being a thatcher, a duster, or a spackler. So, if the agent learns the verbs *to thatch*, *to dust*, and *to spackle* along with the concepts THATCH-EVENT, DUST-EVENT, and SPACKLE-EVENT, it should not associate those concepts with any resulting states. Since most events are not associated with memorable states, the agent does not explore this possibility as a regular part of learning a new event. After all, it is essential to avoid modeling agents who try to be too clever and annoy their human partners by asking too many questions about rare eventualities.

Finalize New Lexical Sense When an agent learns a new concept due to an unknown word or expression, the lexical sense for that word or expression is modified so that it uses the new concept. For example, whereas *systemic sclerosis* was initially mapped to EVENT during language understanding, at this point it gets remapped to SYSTEMIC-SCLEROSIS.

Normalize Event Sequences There are many ways a speaker can present the same information. For example, when teaching a sequence of actions:

- Events can be presented with or without obligative modality.
  - First you grind the coffee beans.
  - First you <u>have to</u> grind the coffee beans.
- Events can be presented in the declarative or the imperative mood.
  - First, you grind the coffee beans.
  - First, grind the coffee beans.
- Indicators of a sequence can be overt or implicit
  - <u>First</u> you grind the coffee beans, <u>then</u> you pour boiling water over them, <u>and then</u> you let it steep for four minutes.
  - You grind the coffee beans, pour boiling water over them, and let it steep for four minutes.
- There can be one or multiple instances of the same kind of modality.
  - If it <u>seems like</u> a patient has achalasia, then . . .
  - If it <u>seems like</u> a patient <u>might</u> have achalasia, then. . . .
- Events can be presented with or without an agent.
  - First you have to grind the coffee beans.
  - First the coffee beans must be ground.

- States can be presented with or without an explanation of how you achieve them.

  - Place ground coffee into a carafe. [What if you have whole beans?]

- Events can be presented in their actual temporal sequence or in a different order.

  - Grind coffee and put it into a carafe.
  - Put coffee into a carafe but first be sure to grind it.

When the agent is learning about how events in the world play out, it needs to normalize such linguistic paraphrases into the intended sequences of actions by stripping obligative modality, pruning out redundant modalities, interpreting *you* as generic, and ordering events temporally. In addition, the agent needs to try to detect and fill in lacunae in the descriptions. For example:

- Property values can be expressed imprecisely: be sure to grind the coffee appropriately for a French press. [How fine is that?]

- Some events in a sequence can be left out if the teacher assumes them to be obvious: if coffee beans are not ground, you have to grind them before making coffee.

- Reasons for events can be left out if the teacher assumes them to be obvious: the water needs to be boiling because cold water cannot quickly extract flavor from coffee grounds (cold brew coffee must steep a lot longer than coffee made with boiling water).

- Methods of detecting the success or failure of component processes, or the script overall, might not be made clear: if the resulting "coffee" turns out to be barely flavored water, reasons might be that the beans were ground too coarsely, the ratio of coffee to water was off, the water was not hot enough, or the brewing time was too short.

In preparation for modeling lacunae detection, it is useful to read documents that describe operating procedures in any domain, such as the troubleshooting instructions for an appliance or vehicle. For novices, these can be remarkably opaque, which reflects the inability of the writers of those documents to adequately mindread their audience.

Learn Property Values Ontological concepts mean what their property values say they mean. A well-described concept differs from its parents and siblings based on property values. However, for practical reasons, at any moment in time, a concept can be insufficiently described, inheriting property values that should have been locally modified. For example, the concept for PENGUIN should indicate that, unlike its parent, BIRD, it is not the AGENT-OF FLY. If the concept description were to lack this information, it would be underspecified. The "Learn property values" function is responsible for the agent's improvement of property-value descriptions of known concepts and its acquisition of property-value descriptions for newly learned concepts. It has to answer for a lot of eventualities, some of which were illustrated by the examples in part 1 of this chapter.

- The systemic sclerosis example (section 7.1) showed the most basic case of learning property values from TMRs.
- The virtual patient example (section 7.2) showed an instance of belief revision, by which the value of the property IS-A changed based on new information. Specifically, HELLER-MYOTOMY was re-anchored from MEDICAL-PROCEDURE to SURGICAL-PROCEDURE.
- During data-driven learning (section 7.3), different examples can suggest different values for a given property, and this information needs to be distilled into what the agent should actually learn.
- The multimodal learning example (section 7.4) showed the reasoning needed to learn a script, which is the sequenced fillers of the SUBEVENTS property of an EVENT.
- The driving example (section 7.5) showed how certain kinds of property values triggered the need for additional reasoning. Specifically, the agent needed to:
  - apply properties of one ontological concept to a coreferential concept being learned (*A stop sign is an octagonal red sign with white letters*);
  - collate different possible values of a scalar attribute into a learnable value (*It is usually about 30 inches across. But it can also be 35 inches wide.*); and
  - translate qualitative attributes into quantitative ones (converting *at/beside a stop sign* into a numerical distance).

All of these reasoning capabilities are provided by procedures in the function "Learn property values," which are as generic as possible but as specific as necessary. Expanding the agent's property-learning capabilities involves increasing the number of eventualities it can treat.

Infer Relations Within a discourse, all adjacent propositions are some-how related, even if the relation is that the topic just switched. Sometimes the relation is expressed linguistically, as by a conjunction (e.g., *and, but, because*), but often it is implicit. For example, *people eat <u>because</u> they're hungry* (7.1), and *<u>if</u> you don't study <u>then</u> you will fail* (7.2).

> (7.1) I'm going to grab an ice cream—I'm famished.
> (7.2) You don't study, you fail.   ; *This requires a particular intonation*

Moreover, even if the discourse relation is overt, it can be expressed using a polysemous conjunction. For example, *and* can indicate an ordered sequence of events (7.3), a causal chain (7.4), related unordered events (7.5), contrastive events (7.6), or an event and its elaboration (7.7).

> (7.3) Scoop ground coffee into carafe, fill with boiling water, stir gently, brew for three minutes, plunge, and pour.
> (7.4) He tripped and fell.
> (7.5) Their toddler whines, cries, and has tantrums—she's a real handful.
> (7.6) My neighbor is great at painting and I'm not.
> (7.7) It's important to keep in touch. And don't worry about bothering me.

The issue of missing and underspecified discourse relations is well known. Computational accounts have primarily involved supervised machine learning that is trained on inventories of relations that have proven too fine-grained even for people to effectively manipulate, no less machines.[25] LEIAs, for their part, already have some initial capabilities along these lines.

They first attempt to infer implicit discourse relations during stage 5 of language understanding, Extended Semantics. If they are successful, those relations are made explicit in the TMR that serves as input to ontology learning and there is no problem. However, if the TMR does not indicate how the meaning of a proposition is linked to the meaning of any previous propositions in the context, or if it is linked to a previous proposition using the vague DISCOURSE-RELATION, then the agent must decide whether or not to explore the relationship further in service of ontology learning. For example, if a human tells its furniture-building robot assistant "Hold the seat, I'll screw it in," this is not a sequence of actions—the agent has to keep holding the seat while the person screws it into the chair frame.

The question is, how can the agent detect when the relationship between events needs to be further specified and then figure out which one is needed? Different opportunities will be available in different kinds of applications. In multimodal applications—as when a robotic LEIA is learning through a combination of processing language and performing physical actions—the temporal correlation of events will play out in the real world, which means that the reasoning can be informed by a combination of language, vision, and even proprioception. In corpus-based learning, the agent will look for multiple occurrences of the same sequences of events, some of which might include the semantic relation between them. If such texts are found, the relation will be learned as part of "Learn property values" rather than "Infer missing relations." Finally, in dialog applications, the agent can ask its human partner about inferred relations, but this needs to be tightly constrained—for example, to cover only cases when the agent cannot determine how to include a new event in a script.

Select Learning Move   The final task in any pass through the learning algorithm is to decide what to do with candidate lexical senses and/or ontological knowledge. In order to select "Record knowledge," the agent has to assess that the candidate knowledge is specific and confident enough to meet the quality threshold required by the application. That could be different, for example, in corpus-based learning about general topics versus dialog-based learning about a critical domain. As for the other options—wait and see, ask a question, launch (or continue) data-driven learning, and cancel learning— their frequency and triggering conditions also vary across applications, as illustrated by the examples in part 1. This general description of "Select learning move" necessarily lacks detail because all of the salient heuristics informing the decision-making are specific to individual applications.

7.3   Final Thoughts on Learning

This chapter has presented a human-inspired model of learning that covers a lot of eventualities in a computer-tractable way that is suitable for LEIAs. Like any model, it reflects choices that could be made differently. However, what is irrefutable is that meaning-oriented learning does involve all of the complexity that the content of this chapter has made clear, and more. Simply dumping a newly encountered word into a lexicon and a corresponding concept into an ontology is not learning in a sense that is relevant for LEIAs.

In order to carry out high-quality learning through reading and dialog, LEIAs need:

1. specialized knowledge about what it means to be a learner: what to learn, in what order, how to assess the quality and confidence of candidate

knowledge, and how to accomplish this with the help of a human collaborator and data-driven tools, such as advanced data analytics and LLMs;

2. specialized knowledge for reasoning about incoming information, as illustrated by the lexicon- and ontology-learning functions discussed throughout the chapter;

3. advanced language understanding capabilities, allowing unconstrained language to serve as input to learning; the field's past experience has shown that people cannot be taught to use simplified language to aid machine applications;[26] and

4. substantial lexical and ontological bootstrapping knowledge, both static and procedural.

But learning through language is clearly not enough. As multisensory agents, LEIAs need to learn about the world using all of their senses. This will result in both ontology enhancement and the enhancement of the appropriate sense-specific knowledge base: the opticon, the hapticon, the physiocon, and so on (cf. section 3.4). Acquiring multisensory ontological knowledge requires collaboration with specialists in domains such as vision recognition, robotic tactile perception, and eye tracking. It is noteworthy that, although manual image-annotation efforts are underway—which could, in principle, form the basis of vision-oriented ontology enhancement—they are largely proprietary.[27]

Learning is never final. The fact that the knowledge learned by a LEIA is at all times incomplete is a feature, not a bug: learning takes time, and a teacher might have more than one training session planned. The important thing is that subsequent learning does not start from scratch since the results of learning are incorporated into the various components of the LEIA's long-term memory to be used during subsequent learning sessions.

In general, we envision learning of the kind described here as a long-term, gradual, increasingly automated method of overcoming the so-called *knowledge bottleneck*. Accordingly, the LEIA is geared toward learning not only previously unknown information but also ever more comprehensive and correct knowledge about entities it already knows about.

Automatic learning by LEIAs is part of the solution to endowing them with broad, deep, high-quality knowledge. The other part is knowledge engineering by humans, which is the topic of chapter 9.

8 Explaining

To earn people's trust, AI systems need to be able to explain their performance. But what does *explain* mean, and what counts as a sufficient explanation? Explanation is one of Marvin Minsky's (2006) "suitcase" words, which are words into which people pack multiple meanings. For example, a doctor's explanation of why you should take your medicine can involve medically sophisticated causal chains, non-specialist-oriented causal chains, projections about what will happen if you don't, population-level statistics about the benefits of the drug, and so on.

Recent years have witnessed an avalanche of publications on trust and explainability in AI, viewed from scientific, technological, philosophical, ethical, and societal angles. But how can one explain data-driven systems that are in principle not explainable? By changing the definition of explanation. In data-driven AI, "explanations" do not attempt to convey how the system works or why it produced the results it did; instead, they provide corroborating evidence for system results that is disjoint from how those results were derived. So, earning people's trust in such systems then becomes a matter of convincing them that non-explanatory corroboration is a reliable measure of the system's competence. While this rejigging might satisfy some end users of some applications, it seems unlikely that it will satisfy individuals who are responsible for outcomes in high-stakes domains.

We believe that for AI systems to be truly explainable, they must be anchored in the kinds of knowledge we have been describing throughout this book. To recap some key features of this knowledge:

- It must be both interpretable by people and optimized for machine reasoning.
- It must include computational cognitive models of the world (ontology), language, and the agent's knowledge of self and others: its goals

and plans, biases, preferences, reasoning methods, and more. Notably, the knowledge should include typical causal chains of events since they provide the best kind of explanation.

• It must include knowledge about how to derive correlational explanations for cases when causal explanations are unavailable.

• It must support processors that translate input data—language, visual inputs, and so on—into knowledge structures that feed machine reasoning.

Agents operating with such knowledge will emulate human behavior in the tradition of folk psychology.[1] For example, a human physician can decide to prescribe a particular medication based primarily on causal information about how it works, but also taking into account population-based statistics about its efficacy and side-effects. Explanatory AI systems must emulate this behavior. Just as a physician can explain to a patient that population-based statistics do not predict how he or she will respond to a medication, so, too, must a clinically oriented AI system.

Explaining recommendations is important and has been a focus of attention for data-driven AI. However, LEIAs need to explain much more than recommendations since they are far more than recommendation systems: they are collaborators that will function in many ways and learn over time through their interactions with people. For such collaborations to be successful, the people involved will need to understand many things about their LEIA teammates—what they know, whether they have successfully learned new information and skills, what they have perceived and done when operating independently, what they plan to do now and why, and so on. Explanation is the window into this inner world of LEIAs, and the microtheory of explanation needs to model it all.

The explanatory potential of the LEIA ecosystem is enhanced by a suite of visualization tools. All LEIA applications that address specialist domains are grounded in models that are (a) available for evaluation by domain experts; (b) open to parameterization, to accommodate different expert opinions; and (c) inspectable by a wide variety of stakeholders: system users, domain experts, educators, funders, investors, and beyond. Both the static and dynamic aspects of these models can be viewed thanks to visualization strategies that will be presented later in this chapter.

Explanation-Oriented Opportunities in the LEIAs' Ecosystem

LEIAs can explain

- what they perceive and how they interpret it;
- their reasoning and decision-making;
- their actions;
- their knowledge of language and the world; and
- what they have learned in a given learning session.

Visualization tools allow developers to demonstrate LEIAs' knowledge and processing to a broad array of stakeholders.

8.1 LEIAs as Social Agents That Explain

LEIAs are social agents that can play various roles in their interactions with people. Some such pairs of roles are shown in table 8.1. In these different roles, LEIAs need to explain different things. For example:

- When a human and a LEIA are collaborating on a task in the same space, they need to either follow predefined rules for their respective roles in the team or negotiate their roles in real time. The latter can require the LEIA to explain its understanding of the plan, everyone's role in it, which step they are carrying out at a given moment, and so on. The collaboration might involve misunderstandings, as can occur between people. If the agent does something unexpected, the human can troubleshoot by asking questions like why the agent did something, what it thought someone said, or what it thought it was asked to do.

Table 8.1
Some roles that humans and LEIAs can play in application systems.

Human Roles	Corresponding LEIA Roles
Live collaborator	Live collaborator
Remote collaborator	Remote collaborator
Teacher	Student
Student	Tutor
Student	Simulated social role (e.g., virtual patient)
Recommendation system user	Recommendation system

- When an agent is a remote collaborator—for example, in the case of an unmanned vehicle—the human must be able to check what the agent is perceiving in its environment, what actions it is taking, the rationale for those actions, and its current plan.

- When a human is teaching a LEIA, the quality and efficiency of that learning depend on the human's understanding of what the agent already knows and whether it has correctly learned the new material.

- When a human is a student, the LEIA can play a particular social role that the human must learn to interact with, such as a virtual patient. Depending on the agent's role, it can need to explain its experiences, its actions, and/or its decision-making.

- When a human is a student, the LEIA can also serve as a tutor in a training environment. As a tutor, the agent needs to not only offer situation-specific advice, flags, and explanations but also put that information into a larger context that serves teaching goals.

- When a LEIA is used as an assistant to a human decision-maker, it provides recommendations, warning flags, and reminders. The agent must be prepared to explain each such move at varying levels of detail.

LEIAs provide explanations under two conditions: when system users ask for them and as a side effect of generating recommendations and warnings. When users ask for explanations, this is a dialog interaction that plays out in the normal way (cf. chapter 6): The agent recognizes the request for explanation as an instance of a particular ontological concept, it instantiates that concept's adjacency pair, and it follows the algorithm recorded in the latter to formulate a response. For example, if I think that the agent misunderstood who I was talking about, I can ask, *Who do you think I was referring to?* The LEIA will recognize this as an instance of the concept REQUEST-ID-OF-REFERENT, whose adjacency pair is EXPLAIN-ID-OF-REFERENT. The latter contains the algorithm that guides the agent in searching its memory for the identity of that individual and conveying it through language. When the LEIA generates an explanation as a side-effect of issuing a recommendation or warning, the default explanation is brief and intended to cover what most users are likely to want to know. If users do not want such explanations, they can turn them off. If they want deeper explanations, they can request them through dialog.

A LEIA's status as a social agent is key to its ability to provide satisfactory explanations. If its initial explanation does not provide exactly what the person wanted to know in terms that were fully understandable, the person can ask follow-up questions. This means that LEIAs will be able to provide useful explanations long before they achieve humanlike sophistication.

There is nothing simple about fashioning an explanation, even after the knowledge prerequisites for it have been met. Consider the example of doctors explaining relevant aspects of clinical medicine to patients. The task has two parts: deciding what to say and how to say it. Both of these depend not only on medical and clinical knowledge but also on the salient features of individual patients as hypothesized by the doctor, such as their health literacy, their interest in medical details, and their ability to process information based on their physical, mental, and emotional states. Identifying these salient features involves mindreading, also known as mental model ascription.[2] An explanation can be presented in many different ways:

- as a causal chain: *You feel tired because of an iron deficiency*;
- as a counterfactual argument: *If you hadn't stopped taking your medicine you wouldn't be feeling so tired*;
- as an analogy: *Most people find it easier to remember to take their medicine first thing in the morning. You should try that*; or
- using a future-oriented mode of explanation: *If you take your medicine regularly, you should feel more energetic*.

Moreover, explanations are not limited to speech—they can include images, videos, body language, live demonstrations, and any combination of the above. Optimizing the automatic generation of explanations tailored to particular individuals in particular circumstances requires a large program of work in itself. However, as with all other aspects of cognitive modeling, simpler solutions can be useful as we make progress over time.

Modeling explanation, like all cognitive modeling, involves anticipating and preparing for eventualities. A convenient way to organize the model is with reference to the agent's cognitive architecture, anticipating what humans might want explained about the agent's perception, reasoning (specifically, the kind of reasoning carried out in the *Deliberation* module of the LEIA's architecture), action, and knowledge resources. For each anticipated explanation need, the agent is provided with an ontological concept containing methods to detect what needs to be explained and a paired concept that guides the agent in providing the explanation (cf. section 6.2). These pairs of concepts, connected by the ADJACENCY-PAIR relation, are recorded in the ontological subtrees headed by REQUEST-EXPLANATION and PROVIDE-EXPLANATION, shown in table 8.2.

It is the leaf concepts in each subtree that contain the agent's reasoning functions. A pair of expanded subtrees, with their leaf concepts in boldface, is shown in table 8.3.[3] The boldface concepts prepare the agent to detect and respond to questions about what the agent heard, what it thought somebody pointed to, and what it saw.

The agent detects what is being asked about—a concept in the left-hand side of the table—using its natural language understanding system. For example, for the agent to understand that *Come again?* is a request to repeat what one just said (REQUEST-REPEAT-STRING), the lexicon needs to map the English construction *Come again?* to the concept REQUEST-REPEAT-STRING. And for the agent to be able to carry out this request, the ontology needs to contain the procedural knowledge, recorded in REPEAT-STRING, to guide its reasoning and action.

Table 8.2
The paired ontological subtrees involving explanation, unexpanded.

– REQUEST-EXPLANATION	– PROVIDE-EXPLANATION
+ REQUEST-INFO-AGENT-PERCEPTION	+ EXPLAIN-AGENT-PERCEPTION
+ REQUEST-INFO-AGENT-ACTION	+ EXPLAIN-AGENT-ACTION
+ REQUEST-INFO-AGENT-KNOWLEDGE	+ EXPLAIN-AGENT-KNOWLEDGE
+ REQUEST-AGENT-REASONING	+ EXPLAIN-AGENT-REASONING

Table 8.3
An example of a pair of expanded subtrees. The leaves are linked using the relation ADJACENCY-PAIR, so when a request on the left is recognized, the concept on the right is instantiated and guides the agent in responding.

– REQUEST-INFO-AGENT-PERCEPTION	– EXPLAIN-AGENT-PERCEPTION
– REQUEST-INFO-PERCEPTION-RECOGNITION	– EXPLAIN-PERCEPTION-RECOGNITION
– **REQUEST-REPEAT-STRING**	– **REPEAT-STRING**
– **REQUEST-POINTED-TO-OBJ**	– **CONVEY-POINTED-TO-OBJ**
– **REQUEST-SEEN-OBJ**	– **CONVEY-SEEN-OBJ**

8.2   Explaining Perception and Action

When humans and agents are collaborating in person, questions about perception and action are useful for coordinating joint activities and for troubleshooting when something goes wrong. When agents are operating remotely, questions are more likely to seek information about things the agent perceives and does that are not directly accessible to the human. Table 8.4 provides some examples.

We will consider a couple of examples in more detail. Imagine that a future robotic LEIA is assisting its human at a barbecue. Among the foods on the grill are veggie burgers. From afar, the human yells:

Human: Turn the veggie burgers!

Robotic LEIA: [Generates a TMR that uses the concept ROTATE to analyze the ambiguous word *turn*, which can mean ROTATE or FLIP-OVER.] OK.
                   [It rotates the veggie burgers.]

Human: [Comes to the grill a couple of minutes later to find veggie burgers that are burnt on the bottom and uncooked on the top.] Why didn't you turn them?!

Robotic LEIA: I did.

Human: What do you think I meant by *turn them*?

Robotic LEIA: Rotate them.

This misunderstanding occurred because the agent selected the wrong sense of the verb *turn*, which can mean either ROTATE or FLIP-OVER. (Note that *Turn the pizza* would mean to rotate it.) This example illustrates a well-known

Table 8.4
Examples of requests for explanation concerning perception and action.

Perception	Action
– What do you think I said?	– What are you doing?
– What do you think I meant?	– Why are you doing that?
– Who/what do you think I pointed to?	– Did you [do something]?
– Who/what do you think I was referring to?	– Where are you going?
– What do you see?	– What are you planning to do next?
– What do you hear?	
– What just happened?	

source of miscommunication: people not realizing that what they are saying is ambiguous or that the addressee doesn't have the world knowledge to understand what is meant.

As another example, consider a situation in which a robotic LEIA is assisting a human in a large room full of human and robotic workers. At some point in a longer exchange, this happens:

Human: Go ask her to come help us.

Robotic LEIA: OK.

              [The LEIA sets off in a particular direction.]

Human: [Watching the LEIA go in the wrong direction]
              Wait, where are you going?

Robotic LEIA: To get Erica.

Human: We don't need her, we need Judy.

This illustrates another possible source of misunderstandings between people: misidentifying the individual referred to by a pronoun.

Perception- and action-oriented explanations share two key similarities: (1) they involve questions that can be answered using knowledge stored in the situation model and (2) answering them is straightforward—the agent need not engage in extensive decision-making about content selection, depth of description, or linguistic formulation. The important thing is for the agent to understand what is being asked of it, which is enabled by the lexicon, where linguistic constructions used to ask questions are mapped to the associated ontological concepts. If the agent doesn't understand what is being asked of it, then it can ask for clarification in the usual way.

8.3   Explaining Knowledge

When humans are teaching or collaborating with a LEIA, they need to understand what the agent already knows and whether it has successfully learned what it was taught. Queries can involve the lexicon, the ontology, and episodic memory. For example:

- Lexicon: *Do you know what upend means? Do you know what it means to upend something? What meanings of the verb upend do you know?*
- Ontology: *Are BoTox injections painful? What is needed to diagnose achalasia? What are the most common colors of cars?* [And, as a follow-up to any of the above] *How do you know?*

- Episodic Memory: *Who performed your Heller Myotomy? When did you finish building this chair? What has Dr. Smith done so far in treating Mrs. Robinson?* [And, as a follow-up] *How do you know?*

As with explaining perception, explaining knowledge requires language understanding to identify what is being asked and procedural ontological knowledge to guide the agent in responding. Specific kinds of requests are recognized as instances of concepts in the ontological subtree REQUEST-INFO-AGENT-KNOWLEDGE, and the algorithms guiding the agent in responding are recorded in adjacency pairs in the subtree EXPLAIN-KNOWLEDGE (cf. table 8.2).

When explaining their knowledge, LEIAs generate the kinds of explanations we think people want when asking different kinds of questions. For example, the answer to yes-no questions is just *yes* or *no* or some paraphrase of them, like *I do* or *I don't*. If I ask an agent *Do you know what a stethoscope is?* and it replies *Yes*, I will trust that it knows the right meaning and will move on. I don't want the agent to habitually elaborate, saying something like *Yes. I think it means a medical instrument for listening to someone's heartbeat or breathing.* Of course, it is possible that the agent knows a different meaning than the one I had in mind, but this unlikely situation does not justify creating agents that are annoyingly verbose. By contrast, if the agent knows more than one meaning of the word, it responds variously based on how many senses it knows. If it knows just two senses, it responds *I know two senses, which mean X and Y*. If it knows three or more senses, it says *I know # senses of that word* and waits to see if the human wants more information. In short, the model of explanation includes normal expectations about how people behave and what they bring to the table when collaborating with agents. If an agent's initial, minimalistic response is not enough, the human can ask a follow-up question as a matter of course.[4]

Answering questions about the content of the ontology depends on what, exactly, is being asked about. There are different algorithms for answering questions about ontological property values, questions about scripts, and broad questions whose answers could involve a variety of ontology elements. Starting with questions about property values, some eventualities are as follows:

- The question can ask about a property that has only a *value* facet, such as the DEFINITION, so the answer is straightforward. For example, the agent will answer the question *What is an EGD?* by generating the filler of the concept's DEFINITION field: *A diagnostic procedure involving examination of the lumen of the esophagus, stomach and duodenum using an endoscope.*

- The question can ask about a property's *default* value. If there is one, the agent reports it; if not, it reports the *sem* value. For example, Q: *What are the most common colors of cars?* A: *White, black, silver, and gray.*
- The question can ask about a property that is defined for multiple facets, in which case the agent needs to incorporate the different values into a fluent English sentence that indicates their status. For example, Q: *Who can perform surgery?* A: *Most commonly, a surgeon, but in some cases, a doctor.*
- The question can be a non sequitur based on the agent's ontological knowledge. For example, Q: *How tall is a snowstorm?* A: *Height isn't defined for snowstorms.*

As regards ontological scripts, questions can be generic or specific, and fielding them can be simple or difficult based on what is asked and how complex the script is. A generic question about a script that contains many subevents with extensive optionality and variability—for example, *Tell me about GERD*—is harder to answer than a specific question about what action comes next in a script that contains only a half dozen strictly ordered subevents—for example, *What do you do after you grind the coffee?*

The point of departure for LEIAs in answering open-ended questions about complex scripts is the script's definition field. For example, the definition of GERD is *GERD is a disease that occurs when acid from the stomach flows back into the esophagus and irritates its lining.* The next layer of explanation leverages the fact that scripts are organized hierarchically, with the nested subevents also having definitions. These can be strung together, with surface smoothing by the language generation system, to explain how a script works. Of course, it is also possible to avoid using the definition fields at all and, instead, construct explanations on the basis of the concept descriptions themselves, using the LEIA's language generation capabilities. This is needed for scripts that the agent learns independently since they do not have explanatory metadata. It is an empirical question in which other situations this processing-heavy approach to explaining a script might be justified.

It is important to underscore that LEIAs are not competing with large language models (LLMs) in answering open-ended questions about the world. When LEIAs field questions, they consult their internal knowledge bases, which contain less information than the training datasets used by LLMs. So, when asked how to make coffee, LEIAs build their response on the basis of their ontological script for making coffee, not unvetted descriptions of coffee-making extracted from uninterpreted texts.

A noteworthy complication of script-oriented questions is that they are often elliptical—that is, they ask for information that is more specific than is obvious from the surface form of the question. For example, the question *What do waiters do?* is a paraphrase of *Tell me the set of events in which waiters typically participate and their role in each of them.* Agents need to infer the specific meanings of such questions during language understanding. For this example, they need to recognize that the construction *What do Xs do?* has a special meaning if Xs is generic and refers to a social role, and *do* is in the present tense, simple aspect. This question is mapped to REQUEST-INFO-SOCIAL-ROLE, whose adjacency pair, DESCRIBE-SOCIAL-ROLE holds the algorithm for formulating the response.

To generalize, in keeping with expectation-oriented modeling, the kinds of script-based questions an agent must be prepared to field include:

- Questions about who or what fills a particular case-role in an event. For example, Q: *Who gives you advice about wine in a restaurant?* A: *The sommelier.*
- Questions about the next event in the sequence. For example, Q: *What do you do after the waiter takes your order?* A: *You wait for your food to be served.*
- Questions about the role of someone or something in the script overall. For example, Q: *What do waiters do?* A: *They explain the menu, take customers' orders, serve food,* and so on.
- Questions that ask for a description of the script overall. For example, Q: *How do you make coffee?* A: *First you set the water to boil, then you grind the coffee beans,* and so on.

Apart from scripts, agents need to be able to field broad questions about ontological knowledge such as *Tell me about penguins.* The basic algorithm is as follows: Indicate the concept's parent; report locally defined property values, which differentiate the child from its parent; if the concept has any subclasses, name them; and if the concept is a script—that is, if it has fillers of the SUBEVENTS slot—launch the EXPLAIN-SCRIPT function. For the penguin example, this will result in: *Penguins are a type of bird. They are black and white, they weigh between 3 and 35 kilograms, they are between 30 and 120 centimeters tall, and they don't fly.* Other clauses in the algorithm for explaining ontological knowledge anticipate requests for further information, such as *What else do you know about penguins?*, and requests that the agent explain what it has learned during a teaching session, such as *Tell me what you now understand about penguins.* In modeling the agent's explanation capabilities, the first priorities are clarity and accuracy, with the smoothness of the language formulation being, at present, less important. This is in contrast to LLMs, which excel at smoothness while having no control over accuracy.

Everything described so far orients around preparing agents to provide specific kinds of answers to anticipated kinds of questions. None of this requires the agent to mindread its interlocutor—that is, to try to figure out why the person is asking the question, what background knowledge he or she already has, and so on. Enabling such mindreading *is* possible and would give the agent more sophisticated explanatory power. However, before undertaking such modeling with respect to explaining knowledge—which is all that we are talking about in this section—we must assess how useful that would be and assign it an associated priority in the overall program of LEIA development.

For narrowly focused questions, mindreading is hardly needed except for choosing which words to use to convey certain information—for example, whether to use or avoid technical terms. The real need for mindreading involves open-ended questions, for which different kinds of answers are appropriate for children, non-specialist adults, domain specialists, and so on. For now, such parameterization is not a high priority for the same reasons described earlier: the person can ask follow-up questions as needed, so the initial explanation need not be perfect, and we expect people to behave in reasonable ways and ask appropriate questions. It would not make sense to ask a LEIA that has significant expertise in clinical medicine a question like *What do you know about health care?* Even a human would balk at such a question.

Turning, finally, to episodic knowledge, explanation requests and responses are very similar to those involving the ontology. The most noteworthy difference between explaining ontological and episodic knowledge involves the fact that episodic knowledge actually comprises two different things: information relating to the agent and its human collaborators, and information about real-world entities outside of the agent's world. By definition, the agent has full knowledge about its private experiences. By contrast, when it comes to public information, it can lack episodic knowledge just like it can lack ontological knowledge. However, recall what we are talking about: explaining *the agent's* knowledge. This is quite different from casting LEIAs as all-purpose question-answering systems, which they are not.

People can ask LEIAs questions about ontology or episodic knowledge either to refresh their memory or because they don't know the answer. If they are just refreshing their memory and the response sounds right, then no explanation is needed. By contrast, if the answer does not sound right, or if the information is new, then they might want to validate its veracity. To do so, they might ask questions like *How do you know?* or *What exactly did you read/hear/find?* In some cases, the agent will have recorded the source of information as metadata, as when it engages in learning by reading or is being instructed by a particular human. In cases when the source of information is not known, the agent could be instructed to search a corpus for corroborating information that includes a source; but this goes beyond the basic functionality of enabling agents to explain their current state of knowledge.

This section has described how a LEIA can explain the content of its lexicon, ontology, and episodic memory. This does not exhaust its knowledge. Another aspect of its knowledge is that which underlies its reasoning, to which we now turn.

8.4 Explaining Reasoning

According to the AGENT-FUNCTIONING-FLOW script that implements the LEIA's architecture (cf. section 3.2.4), agent action is triggered in four ways:

1. The action can be the output of a specific decision function. For example, when the agent is serving as an advisor, if it detects a user error, it issues a warning flag.
2. The action can be the ADJACENCY-PAIR of the previous event. For example, when a person asks a question, the agent answers it.

3. The action can be triggered by a daemon (a standing goal). For example, a particular virtual patient might need to know if procedures are painful before agreeing to them. If the doctor recommends a procedure for which the patient lacks information about pain, it posts the goal of tracking down that information.

4. The action can be the next step in the current plan on the agent's agenda, launched when the previous step is completed. For example, if a robotic LEIA is building a chair, then after it attaches the first leg, it undertakes to attach the second leg, and then the third, and the fourth.

In all cases, the agent knows why it chose a given action. Formally, this is recorded as metadata with the remembered instance of the event. For cases 2–4, explaining the reasoning is relatively straightforward, apart from some details of language generation. If the agent is asked *Why did you say that?* or *Why are you doing that?* it answers:

2a. Because [a brief description of the first event in the adjacency pair]. For example, Q: *Why did you say that?* A: *Because Joe asked me where I was going.*

3a. Because [a brief description of the daemon]. For example, Q: *Why did you ask that question?* A: *Because I need to know if procedures are painful before agreeing to them.*

4a. It's the next step in [plan name]. For example, Q: *Why are you doing that?* A: *It's the next step in building a chair.*

By contrast, explaining decision functions, the first case above, is more complicated since those decisions can involve not only agent actions but also recommendations, advice, warnings, and so on. Examples of what a person might want to know about decisions include, among many others:

- Why did you do that?
- Why did you recommend that?
- Why do you think that what I was planning to do is wrong?
- What did you base your recommendation on?
- Was your recommendation informed by machine learning?
- Would any additional information be useful in making this decision?
- How sure are you of this recommendation?
- Are you sure this is best?
- Why is [option X] better than [option Y]?

> The most important point about preparing LEIAs to explain their decision-making is that **all of the information they need is either recorded as knowledge associated with the decision function or is dynamically generated and stored as metadata while the decision is being made.** This means that agents do not need to invent or reconstruct the reasoning behind their decisions if they are asked about them; they need only look up what was already prepared and then package it as a situationally appropriate utterance.

The static and dynamically generated information a LEIA relies on to explain its decision-making is recorded as values of the following metalevel properties:

- EXPL provides a short English explanation of a decision function; it typically includes variable slots.
- CONFIDENCE holds the agent's confidence in the decision, measured on the abstract scale {0,1}.
- RELEVANT-FEATURES holds the list of properties whose values, if known, contribute to the decision.
- CONTRIBUTING-FEATURE-VALUES holds the list of actual property values contributing to the decision in the given context.
- ABSENT-FEATURE-VALUES holds the list of properties for which the decision function needs values that are unknown in the given context.
- ROLE-OF-ML conveys the role of data-driven methods in the decision, apart from its role in perception processing, which is considered separately; the role of ML affects the agent's confidence in the decision.
- COMPARISON-OF-OPTIONS holds the result of comparing decision options that are above a quality threshold.
- IDIOSYNCRATIC-DECISION-TRACE allows for any other aspects of the decision function to be prepared in advance as an explanation; for example, a numerical calculation that contributes to a decision can be described in detail in plain English.

We will describe the use of these features on the example DETECT-JUMPING-TO-CONCLUSIONS-DIAGNOSIS, which is used by agents that are serving as medical tutors or advisors. This algorithm detects whether a diagnosis posited by a system user is clinically appropriate. Here we focus on how this algorithm's content supports explanation; how the algorithm is used in a medical application system is described in section 8.5.

DETECT-JUMPING-TO-CONCLUSIONS-DIAGNOSIS

    DEFINITION   This procedure detects if a doctor has sufficient evidence to make a diagnosis.

    AGENT   LEIA

    SUBEVENTS

      TRY: sufficient-grounds-to-diagnose-ok

        EXPL   "Diagnosing [DISEASE] is clinically valid. The relevant feature values are [CONTRIBUTING-FEATURE-VALUES]."

      TRY: no-sufficient-grounds-to-diagnose

        EXPL   "Diagnosing [DISEASE] is not clinically valid. [CONTRIBUTING-FEATURE-VALUES] contribute to making the diagnosis but [ABSENT-FEATURE-VALUES] must also be known."

This algorithm has two conditions. The first one detects situations in which all of the necessary preconditions for diagnosing the disease, which are recorded in the ontology, have been met, so the move is clinically valid. When this condition holds, the agent explains its decision by generating the content of the EXPL field, which includes two dynamically populated variable slots: the name of the disease and the feature values that made the diagnosis valid. The CONFIDENCE in this decision is 1 (fully confident) because the decision involves simply comparing feature values in the ontology with feature values in the dynamically populated patient model.

The second condition covers cases in which necessary preconditions for the diagnosis have not been met. The agent explains this decision as before: by generating the content of the EXPL field, which has a different set of dynamically populated variable slots. The CONFIDENCE in this decision is also 1 for the same reason as above.

The final three features in the list above—ROLE-OF-ML, COMPARISON-OF-OPTIONS, and IDIOSYNCRATIC-DECISION-TRACE—are not applicable for this particular decision function but they are needed for others, such as recommending a particular treatment option.

8.5 An Example: LEIAs Serving as Tutors and Advisors Explain Their Reasoning

Tutoring students and advising professionals have much in common. LEIAs use the same knowledge and reasoning for both but package messages differently for the different audiences. For this overview, we will disregard minor differences between tutoring and advising and treat them as a single capability. But before getting into how LEIAs tutor and advise, we need some background about an important source of errors in human decision-making: cognitive biases.

Cognitive bias is a term used by psychologists to describe distortions in human reasoning that lead to empirically verified, replicable patterns of faulty judgment (Kahneman, 2011). Cognitive biases result from the inadvertent misapplication of necessary human abilities: the ability to simplify complex problems, make decisions despite incomplete information (i.e., decision-making under uncertainty), and generally function under the real-world constraints of limited time, information, and cognitive capacity (cf. Simon's [1957] theory of bounded rationality). Factors that contribute to cognitive biases include, non-exhaustively:[5]

- overreliance on one's personal experience as heuristic evidence;
- the misinterpretation of statistics;
- overuse of intuition over analysis;
- acting from emotion;
- the effects of fatigue;
- considering too few options or alternatives;
- the illusion that the decision-maker has more control over how events will unfold than he or she actually has;
- overestimation of the importance of information that is readily available over information that is not;
- framing a problem too narrowly; and
- not appreciating the interconnectedness of multiple decisions.

Even if one recognizes that cognitive biases could be affecting decision-making, their effects can be difficult to counteract. As Heuer (1999) writes, "Cognitive biases are similar to optical illusions in that the error remains compelling even when one is fully aware of its nature. Awareness of the bias, by itself, does not produce a more accurate perception. Cognitive biases, therefore, are, exceedingly difficult to overcome" (112). When serving as tutors and advisors, LEIAs can offer objective assessments of when a cognitive bias might be at play. This should be more useful than simply reporting potential user errors with no insight into where the persons' reasoning might have gone wrong.

Returning to LEIAs, they know how to tutor and advise on the basis of the ontological subtree headed by the concept TUTORING-AND-ADVISING. This concept indicates that tutoring and advising comprise two kinds of actions: (1) evaluating user moves and flagging problems, and (2) answering questions.

TUTORING-AND-ADVISING
   DEFINITION   Tutoring and advising involves evaluating user moves, flagging
                    errors, and answering questions.
   AGENT   HUMAN, LEIA
   BENEFICIARY   HUMAN, LEIA
   SUBEVENTS
     EVALUATE-FLAG-USER-MOVE
     RESPOND-TO-REQUEST-INFO

TUTORING-AND-ADVISING is an intermediate node in the ontology that is not ever instantiated as a script, but it contains the knowledge that allows an agent to answer questions like *What is involved in tutoring and advising?*

The first subevent of TUTORING-AND-ADVISING, called EVALUATE-FLAG-USER-MOVE, is a script that has its own subevents, which involve detecting both plain errors and errors resulting from cognitive biases.[6]

EVALUATE-FLAG-USER-MOVE
   DEFINITION   This script evaluates user moves and detects different kinds of
                    errors, including those that might result from cognitive biases.
   AGENT   LEIA
   BENEFICIARY   HUMAN
   SUBEVENTS
     TRY: DETECT-PLAIN-ERROR
     TRY: DETECT-JUMPING-TO-CONCLUSIONS
     TRY: DETECT-FRAMING-SWAY
     TRY: DETECT-SMALL-SAMPLE-BIAS
     TRY: DETECT-BASE-RATE-NEGLECT
     TRY: DETECT-ILLUSION-OF-VALIDITY
     TRY: DETECT-EXPOSURE-EFFECT

In teaching and advising applications, the LEIA evaluates each move by the user to see if it reflects any known kind of mistake. Procedurally, this means the LEIA tests whether each move fulfills the preconditions of any of the SUBEVENTS of EVALUATE-FLAG-USER-MOVE. If the preconditions for detecting any of these error types are met, the agent issues an associated message.

EVALUATE-FLAG-USER-MOVE is an intermediate node in the ontology whose description is useful if someone asks an agent a metalevel question like *What do tutors do?* or *What kinds of mistakes do tutors look out for?* However, in order for the agent to perform as a tutor or advisor in an application, it needs domain-specific knowledge, which is recorded in the appropriate descendant of TUTORING-ADVISING-SCRIPT, for example:

TUTORING-ADVISING-SCRIPT

    TUTORING-ADVISING-**CLINICAL-MED**

    TUTORING-ADVISING-**FURNITURE-ASSEMBLY**

    TUTORING-ADVISING-**DRIVING-A-VEHICLE**

Taking the example of clinical medicine, when a tutoring or advising session for clinical medicine begins, the agent places an instance of TUTORING-ADVISING-**CLINICAL-MED** on its agenda. One of its subevents is EVALUATE-FLAG-**CLINICAL-MED**-MOVE, which holds the knowledge about how to respond specifically to actions in the domain of clinical medicine.

EVALUATE-FLAG-**CLINICAL-MED**-MOVE

DEFINITION	This script evaluates user moves in the domain of clinical medicine and detects different classes of errors, including those that might result from cognitive biases.
AGENT	LEIA
BENEFICIARY	HUMAN

SUBEVENTS

    TRY: DETECT-PLAIN-ERROR-**MED**

    TRY: DETECT-JUMPING-TO-CONCLUSIONS-**MED**

    TRY: DETECT-FRAMING-SWAY-**MED**

    TRY: DETECT-SMALL-SAMPLE-BIAS-**MED**

    TRY: DETECT-BASE-RATE-NEGLECT-**MED**

    TRY: DETECT-ILLUSION-OF-VALIDITY-**MED**

    TRY: DETECT-EXPOSURE-EFFECT-**MED**

For purposes of illustration, we will describe two of these subevents: the one that detects jumping to conclusions, which we introduced in passing earlier; and the one that detects presenting information using a framing sway, which involves phrasing it in a way that could subtly influence the hearer's response to it.

Agents detect jumping to conclusions by comparing user moves against the preconditions of good practice recorded in the ontology. Good practice is encapsulated in guidelines, agreed upon by clinicians, that inform clinical decision-making. For example, there are guidelines for determining when there is enough evidence to hypothesize or diagnose a disease, and when it is justified to recommend tests and interventions. Table 8.5 shows some of the preconditions of good practice recorded in the ontological description of the disease achalasia. For readability, the fillers are described using English strings rather than meaning representations written in the ontological metalanguage.

Using this information, agents functioning as tutors and advisors can detect and flag if users are making moves that are not yet clinically justified, and they can answer questions like *Can a diagnosis be made yet?*

A knowledge-engineering aside. Tables like 8.5 served as the common ground between knowledge engineers and physician educators during development of the Maryland Virtual Patient system, and they are useful in presenting cognitive models to students, educators, and other stakeholders.

Although we already presented the algorithm for detecting jumping to conclusions about diagnoses, we repeat it here for easy comparison with table 8.5, in order to emphasize how agents leverage knowledge while evaluating decision functions.

Table 8.5
Some of the preconditions of good practice related to the disease achalasia.

Property	English gloss of filler
SUFFICIENT-GROUNDS-TO-SUSPECT	Dysphagia to solids and liquids or regurgitation
SUFFICIENT-GROUNDS-TO-DIAGNOSE	All four of the following conditions: 1. either bird's beak or lower esophageal sphincter pressure > 45 2. aperistalsis 3. either dysphagia or regurgitation 4. negative EGD for cancer
SUFFICIENT-GROUNDS-TO-TREAT	Definitive diagnosis of achalasia
PREFERRED-ACTION-WHEN-DIAGNOSED	Pneumatic dilation or Heller myotomy
REASONABLE-ACTION-WHEN-DIAGNOSED	Administer BoTox

DETECT-JUMPING-TO-CONCLUSIONS-DIAGNOSIS

DEFINITION   This procedure detects if a doctor has sufficient evidence to make a diagnosis.

AGENT   LEIA

SUBEVENTS

TRY: sufficient-grounds-to-diagnose-ok

   EXPL   "Diagnosing [disease] is clinically valid. The relevant feature values are [contributing-feature-values]."

TRY: no-sufficient-grounds-to-diagnose

   EXPL   "Diagnosing [disease] is not clinically valid. [contributing-feature-values] contribute to making the diagnosis but [absent-feature-values] must also be known."

If a user diagnoses a disease, the agent evaluates the move against its knowledge of good clinical practices. If the move is valid, then the agent simply records it, including why it was justified, in case the agent is asked about it later on. The justification is a trace of the preconditions that were fulfilled in order to satisfy the conditions of the function sufficient-grounds-to-diagnose-ok.

If, by contrast, a user move is incorrect, then the agent instantiates the concept FLAG-CLINICAL-MOVE, which is the output of the function no-sufficient-grounds-to-diagnose, and passes to that function the information about which move was erroneous and why.

Consider the following example that shows how a LEIA playing the role of a tutor responds when a clinician in training posits a diagnosis without sufficient evidence.

- The student says to the virtual patient *You have achalasia*.
- The LEIA's language understanding system understands this to mean "DIAGNOSE-DISEASE-1 (THEME ACHALASIA-1) (ACHALASIA-1 (EXPERIENCER PATIENT-1))."
- Since TUTORING-ADVISING-CLINICAL-MED is on the agent's agenda, given the agent's role as a tutor, the agent tests everything it perceives against all of that plan's SUBEVENTS, one of which is EVALUATE-FLAG-CLINICAL-MED-MOVE.
- We assume for this example that the diagnosis matches the second condition, no-sufficient-grounds-to-diagnose, because although the first three conditions of SUFFICIENT-GROUNDS-TO-DIAGNOSE in table 8.5 are fulfilled, the fourth one is not: negative EGD for cancer.
- The LEIA records relevant metadata associated with this decision:

  - CONTRIBUTING-FEATURE-VALUES (in plain English): a bird's beak, aperistalsis, and dysphagia.
  - ABSENT-FEATURE-VALUES (in plain English): negative EGD for cancer.

- The LEIA instantiates FLAG-CLINICAL-MOVE, and its language generation system outputs: "Diagnosing achalasia is not clinically valid. A bird's beak, aperistalsis, and dysphagia contribute to making the diagnosis but a negative EGD for cancer must also be known."[7]

Dedicated algorithms relying on ontological knowledge similar to that in table 8.5 are available to the agent for evaluating whether the preconditions of good practice have been fulfilled for recommending tests and procedures as well.

Which kind of warning the agent generates depends on the type of user, student vs. professional, as well as user preferences about how much information to provide when issuing warnings. For example, in the Maryland Virtual Patient application, students could choose to see no tutoring messages, minimalistic messages (e.g., there is insufficient evidence to diagnose a disease), messages with context-specific information (e.g., what else must be known in order to diagnose the disease in the given patient), or messages with extensive additional information (e.g., all of the different ways of fulfilling the preconditions for diagnosing the given disease). The information-rich option is illustrated by figure 8.4 in section 8.7, which shows all clinically justified grounds for ordering the test called EGD as well as which preconditions were already satisfied at the given point in the given simulation run.

The second example of a flaggable user move that we will consider involves linguistic priming. In clinical scenarios, the way a doctor describes interventions, presents options, and asks questions can impact patients' impressions and their subsequent decision-making. For example,

- If the doctor asks "I imagine your throat hurts, right?," the patient will have a tendency to seek corroborating evidence, even if he or she had not previously noticed any throat pain. This is the *confirmation bias*.

- If the doctor asks "Your pain is very bad, isn't it?," the patient is likely to overestimate the perceived pain, having been primed with a high pain level. This is the *priming effect*.

- If the doctor says, "There is a 20% chance that this procedure will fail," the patient is likely to interpret the procedure more negatively than if the doctor had said, "There's an 80% chance that this will succeed." This is the *framing sway*.

Agents can help doctors to be aware of, and learn to avoid, such formulations by flagging potentially bias-inducing utterances. The detection methods involve recognizing linguistic constructions as particular ontological concepts, which are located in the DETECT-BIASED-LANGUAGE branch of the SPEECH-ACTS subtree of the ontology. Table 8.6 provides examples.

Although the agent can detect utterances with a framing sway, it remains a research issue when it should report them. For example, it is entirely appropriate, and an indication of compassion, for a doctor to say to a patient who has asked for an increase in pain medication, "So, the pain is pretty bad?"

In addition to proactively responding to user moves, agents functioning as tutors and advisors can answer questions in the normal way. The agent interprets the questions as instances of specific ontological concepts using constructions in its lexicon and it looks up the adjacency pair of the relevant concept to determine how to answer. Table 8.7 provides examples.

Table 8.6
Examples of constructions that can lead to biased decision-making.

Example	Associated bias-detection function
I assume you don't eat before bed, right?	DETECT-SEEK-CONFIRMATION-QU
Do you feel a sharp pain in your chest?	DETECT-SUGGESTIVE-YES/NO-QU
Do you have heartburn between 10 and 20 times a week?	DETECT-PRIME-WITH-RANGE-QU
There's a 15% chance this procedure will fail.	DETECT-NEGATIVE-FRAMING-SWAY
There's an 85% chance this procedure will succeed.	DETECT-POSITIVE-FRAMING-SWAY

Table 8.7
Examples of the adjacency pairs for asking questions about clinical moves.

Sample Questions	Concepts Reflecting the Meaning of the Questions	The Adjacency Pairs Guiding the Response
Can I diagnose achalasia yet?	REQUEST-INFO-DIAGNOSIS-POTENTIAL	EVALUATE-DIAGNOSIS-POTENTIAL
Which diagnoses should I be thinking about?	REQUEST-INFO-DISEASE-HYPOTHESIS-POTENTIAL	EVALUATE-DISEASE-HYPOTHESIS-POTENTIAL
Is it OK to order an EGD?	REQUEST-INFO-TEST-ORDERING-POTENTIAL	EVALUATE-TEST-ORDERING-POTENTIAL
Would Heller Myotomy be a reasonable recommendation?	REQUEST-INFO-MED-INTERVENTION-POTENTIAL	EVALUATE-MED-INTERVENTION-POTENTIAL

All of the algorithms guiding the agent's responses use the same kinds of ontological knowledge as illustrated in the above tables. For example, the EVALUATE-DIAGNOSIS-POTENTIAL script evaluates and reports on whether:

- the known feature values of a patient make it possible to diagnose a disease; if so, which one or ones;
- the known feature values of a patient make it possible to hypothesize an as-yet not hypothesized disease;
- the known feature values of a patient are not sufficient to diagnose a hypothesized disease; if so, which features values are missing; and
- the known feature values of a patient make it impossible to diagnose or hypothesize any disease.

If users want more information than is provided by this algorithm, they can ask for it. For example, if, at the given time, no disease can be diagnosed or hypothesized and the agent says so, the human can follow up with a question about related ontological knowledge, such as "What is needed to diagnose [some disease]?" The agent will answer using its methods for explaining ontological knowledge, described in section 8.3.

8.6   How Empirical Contributions to LEIA Operation Affect Explainability

Data analytics and machine learning have important roles to play in explanation-capable systems. For example, visual object recognition, speech recognition, and syntactic parsing can be performed using data-driven methods, and data-driven recommendation systems can inform LEIA decision-making. The question is, how do the unexplainable contributions of data-driven systems affect the overall explainability of LEIA operation? The answer depends on the system module in question.

Perception Recognition is primarily handled by data-driven tools, such as image and speech recognition systems. When an image recognition system recognizes an image as a tree, or a speech recognition system recognizes an utterance as *His bicycle isn't*, those results cannot be explained. In straightforward cases—when the signal is clear and the recognition is confident—the lack of explainability is not a problem. However, when a signal is unclear, incomplete, or ambiguous, Perception Recognition tools perform worse than people. This is because people can use the context and their knowledge of the world to recognize a largely occluded object or a disrupted speech signal, and they can explain how they did it. For practical purposes, the unexplainability of data-driven Perception Recognition is not a problem when the answer is right, but it is a big problem when the answer is not right. Of course, if agents are ever to achieve humanlike capabilities, they need to be able to explain even high-confidence Perception Recognition in terms of feature values—for example, how to distinguish a dog from a cat with reference to features that they can describe and point to.

Perception Interpretation—that is, translating raw recognition output into ontologically grounded meaning representations—also incorporates certain data-driven tools. For example, use of a statistical syntactic parser means that the agent can say what the parse is but not why. This does not seriously undermine the explainability of language understanding since syntactic parsing plays only a supporting role in what is overwhelmingly a semantic and pragmatic process.

Data-driven tools can also inform the reasoning carried out in the *Deliberation* module of the LEIA's architecture. For example, LEIAs could use LLMs to help them to estimate unknown feature values in decision functions and to incorporate population-level statistical evidence into their decision-making. However, since the output of LLMs is not explainable, LEIAs must be prepared to incorporate into their own explanations the role and relative weight of such evidence in their overall decision-making. Hybrid decision-making of this type is a practical approach to getting the best from knowledge-based and data-driven approaches; and making the status of the resulting decisions maximally explainable is key to gaining the trust of human decision-makers.

8.7 Visualizations for Explanation in the Maryland Virtual Patient System

The model of explanation described in this chapter reflects our first attempt at a broad-coverage microtheory of explanation, but this is not the first time that explanation has been a part of LEIA modeling. For the Maryland Virtual Patient (MVP) clinician-training application, a core requirement was explainability to a variety of stakeholders—teachers, non-teaching domain experts, students, developers, and funders. A core strategy for fulfilling this requirement was the use of visualizations, three examples of which we present below by way of illustration.

> **Visualization as Explanation.** The goal of explanation is to provide insight into what is being explained. Language is one way of providing such insight, and visualizations are another. Visualizations are a useful method of explaining knowledge bases, cognitive models, and algorithms, and they are indispensable for providing traces of system processing that are accessible to developers and non-developers alike.

Visualizing the physiological models underlying virtual patients   Modeling human physiology to support dynamic, interactive virtual-patient simulations is not about trying to replicate a human in the box. Instead, a knowledge engineer leads physicians serving as subject matter experts through the process of distilling their extensive knowledge about physiology and clinical practices into the most relevant subset and expressing it in sufficiently formal terms. Not infrequently, specialists are also called on to hypothesize about the unknowable, such as the preclinical (i.e., pre-symptomatic) stage of a disease and the values of physiological properties between the times when tests are run to measure them. Such hypotheses are, by nature, imprecise. However, rather than permit this imprecision to grind agent building to a halt, we proceed in the same way as live clinicians do: by developing *a* model that is reasonable and useful, with no claims that it is the only model possible or that it precisely replicates human functioning.[8]

Certain kinds of diseases can be conveniently divided into conceptual stages, with disease progression being represented as changes to particular physiological properties and patient symptoms over time. Table 8.8 illustrates this using a model of the disease achalasia.[9] Achalasia makes a person's lower esophageal sphincter (LES) hypertensive and reduces the efficacy of esophageal peristalsis, which results in difficulty swallowing and various other symptoms.

The top portion of the table shows how physiological property values change over time (the stages labeled t0–t4) if the disease is left untreated, and the lower portion shows patient symptoms given a particular physiological state. In this model, some features have different values across patients (default values are shown in square brackets, and legal ranges are shown when they are constrained), whereas other features play out the same across patients. The latter does not imply that all human patients are the same with respect to these features. Instead, it reflects the fact that this is a model—by necessity, a simplification—that aims to serve particular

Table 8.8

The portion of the model of achalasia that shows changes in physiological feature values and associated patient symptoms if the disease is left untreated.

	Start	t0	t1	t2	t3	t4
Stage duration (in months)		[12]	[12]	[12]	[12]	[12]
Physiological Properties						
Ratio of relaxing to contracting neurons in the distal esophagus	100/100	80/100	60/100	40/100	20/100	10/100
Basal LES pressure (torr)	0–40 [25]	[start-value]+8	[start-value]+16	[start-value]+24	[start-value]+32	[start-value]+40
Residual LES pressure (torr)	0	8	16	24	32	40
Residual LES diameter (cm)	2	1.5	1.0	.5	0	0
Amplitude of contraction during peristalsis	80	65	40	30	20	10
Efficacy of peristalsis	normal	normal	normal	intermittent peristalsis	aperistalsis	aperistalsis
Diameter of distal esophagus (cm)	2	2.8	3.6	4.2	5	6
Retained esophageal content (on the scale [0,1])	0	0–2 [.1]	.2–4 [.3]	.4–.7 [.55]	.7–1 [.85]	1
Emptying delay (minutes)	0	1	5	10	30	$35{,}000^{10}$
Symptoms						
Difficulty swallowing distal (on the scale [1,4])	0	.1	.5–1 [1]	1–2 [2]	2–3 [3]	3–4 [4]
Do solids stick while swallowing?	no	no	yes	yes	yes	yes
Do liquids stick?	no	no	no	yes	yes	yes
Weight loss (on the scale [0,1])	0	0	0–05 [0]	0–1 [0]	0–15 [.1]	.05–2 [.2]
Chest pain (on the scale [0,1])	0	0	0–3 [.1]	0–5 [.3]	.3–8 [.5]	.5–1 [.7]
Regurgitation (times per month)	0	0	0–4 [0]	0–20 [10]	20–50 [40]	20–100 [70]

pedagogical purposes defined by the physician-educators who collaborated on building the system. They did not believe that adding additional variability to the model would improve the educational experience.

Whereas table 8.8 reflects how achalasia unfolds if there are no interventions, clinical medicine is all about interventions. Interventions are also modeled using tables, but ones with a different semantics. Table 8.9 shows how BoTox injected into the LES works as an intervention for achalasia. BoTox is not a cure for the disease, but it can reduce symptoms for up to a year.

Intervention tables cover both the case when the treatment is given to a previously untreated patient and to a patient whose LES pressure has been changed by other interventions. This highlights the fact that MVP simulations are open-ended, not fixed paths in the style of decision trees.

Table 8.10 shows an example of how treatment plays out in a simulation run of a particular virtual patient, Gladys. Gladys is given BoTox when her LES pressure is 52. The BoTox injection brings her LES pressure down to 32 which, as the basic disease table 8.8 shows, is a normal pressure that does not evoke symptoms. However, the effect of BoTox will wear off over six months, returning Gladys to her original LES pressure of 52.

Table 8.9

The model of how BoTox works as an intervention for achalasia, which allows for great variability across patients.

If a patient is given BoTox when his or her basal LES pressure is	33–40	41–48	49–56	57–64	65+
Then his or her basal LES pressure will initially go down to	4–24 [15]	12–30 [21]	18–36 [27]	24–42 [33]	30–48 [39]
And the effect will wear off over # months	6–18 [12]	6–18 [12]	6–18 [12]	6–18 [12]	6–18 [12]

Table 8.10

An example of how a particular virtual patient, Gladys, will respond to BoTox if it is injected when her LES pressure is 52.

If a patient is given BoTox when his or her basal LES pressure is	33–40	41–48	52	57–64	65+
Then his or her basal LES pressure will initially go down to	4–24 [15]	12–30 [21]	32	24–42 [33]	30–48 [39]
And the effect will wear off over # months	6–18 [12]	6–18 [12]	6	6–18 [12]	6–18 [12]

Since all values in the treatment table are variables, and since the variability covers large scales, virtual patients can play out very differently with respect to their response to BoTox if a user should decide to use it as a treatment in a simulation run. Other available treatments for this disease—the surgical procedure called Heller myotomy and the endoscopic procedure called pneumatic dilation—are modeled using similar tables.

The main point of this discussion is not what, exactly, all of this means medically. The important thing is that this relatively simple model: (a) is fully transparent and, therefore, explainable; (b) is easily extensible and modifiable—that is, it can be augmented to reflect new findings in medicine or different opinions of different domain experts; and (c) is able to generate great variability across the population of virtual patients, thus fulfilling the pedagogical goals for which it was developed.

Visualizing system operation using under-the-hood panes   Another example of visualization-based explanation in the MVP system involves displaying dynamic traces of system operation in what we call under-the-hood panes. The inventory of panes includes the following:

Physiology:	A list of disease-relevant property value pairs, with values being highlighted every time they change during the simulation. This reflects an omniscient view of the patient's physiology.
Interoception:	A list of the virtual patient's perceived symptoms as property-value pairs. Every time a symptom appears or changes, a new entry is posted.
Thoughts:	Dynamically populated traces of the patient's decision functions, rendered in plain English for readability; for example, "I don't know the risks of EGD. I'd better ask about them."
Knowledge Learned:	Traces of words, ontological concepts, and property values of concepts learned through dialog.
TMRs:	Text meaning representations of the virtual patient's interpretations of the user's inputs during the simulated doctor-patient interactions.

In the proof-of-concept MVP system, these panes were presented as side-by-side columns in the lower portion of the computer screen during simulation runs. Screenshots from two relatively self-explanatory panes during a particular simulation run are shown in figure 8.1.[11]

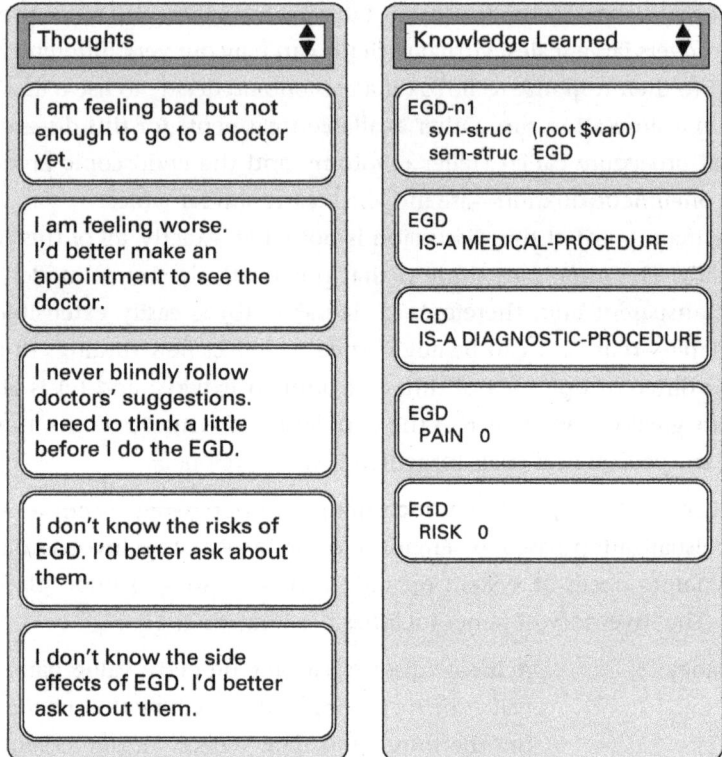

Figure 8.1
Two of the under-the-hood panes of the MVP system during a simulation run.

The *Thoughts* pane shows traces of the patient's decision-making that were generated at different points in the simulation. First it had to decide about going to see the doctor, then about whether to agree to the intervention called EGD. Its decision-making is influenced by its lack of knowledge about the procedure as well as its character traits, which include not blindly trusting doctors and wanting to know specific things about procedures before agreeing to them. The *Knowledge Learned* pane shows traces of what the agent learns about EGD at sequential steps of the interaction. First it recognizes that EGD is a noun that must be added to its lexicon and maps it to a new concept called EGD, which it assumes must be some kind of MEDICAL-PROCEDURE. Then, after asking the doctor (the system user) for more information about the procedure, it learns that it is actually a DIAGNOSTIC-PROCEDURE that is not painful and carries no risk.

The under-the-hood panes of MVP not only show exactly what is happening during the simulation; they also show that the simulation system, although a prototype, is real: its components are modeled in such a way as to be extensible into a deployable system.

Visualizing tutoring content The third kind of explanatory visualization in MVP involves tutoring. The system offers various options regarding when to provide tutoring messages and what to display in them. Tutoring messages can be provided only when the user is about to make a mistake, every time the user makes a major move (orders a test or procedure, hypothesizes a disease, or diagnoses a disease), or not at all. As concerns what to show, the messages range from a minimal right/wrong indicator to full information about the preconditions of good practice related to the move, including which preconditions are currently fulfilled with respect to the given patient. This latter strategy aims to teach by repetition, reinforcing the full cluster of related knowledge each time a move is made. Figure 8.2 shows an example of a tutoring message that appeared in a pop-up window when a user ordered the test called EDG during a simulation run. In the system, the messages were color-coded using green to indicate fulfilled preconditions

Preconditions for EGD

ONE OF:

Suspicion of a mechanical obstruction

Suspicion of GERD

Suspicion of Barrett's esophagus

Suspicion of achalasia (to rule out pseudoachalasia)

ONE OF:

Dysphagia

10% weight loss

Figure 8.2
An example of a tutoring message. Boldface shows preconditions that have been satisfied. The preconditions are conceptually grouped into suspicions about diseases and disorders (the first four) and individual symptoms (the last two). Dysphagia is enough to warrant ordering an EGD.

and red to indicate unfulfilled ones. Here we use boldface for the green ones and plain Latin for the red ones.

To wrap up this discussion of explanation in MVP, a variety of visualization techniques are used to explain both the medical model and system functioning. The goals of these visualizations are to ensure that the system is functioning correctly, to gain the trust of medical educators, and to explain to users and the outside world both the model itself and how it plays out in simulation.

8.8   Explanation as Part of Overall Agent Operation

Although this chapter has focused on explanation, it has not been exclusively *about* explanation. It has been about how explanation plays into the overall operation of LEIAs. The kinds of explanations LEIAs will be called on to provide depend on what role they are playing in a given application. The content of the explanations depends on the LEIAs' static knowledge, their situational knowledge, and the algorithms driving their operation. How they determine what needs to be explained and how to generate an explanation derives from their microtheory of explanation itself.

An important aspect of this microtheory is the acknowledgement that agents need to be prepared to recognize and reason about different kinds of requests for information. There is not, nor can there be, a single question-answering capability since questions like *What do you think I said?* and *Why is it too early to posit this diagnosis?* require completely different reasoning.

One question that might come to mind is, *Couldn't the agent look directly at its codebase to explain what it is doing?* To some degree, yes. Decision functions use ontological property values, and so, at a minimum, the agent can extract those and report that they affected the decision. However, two important points must be kept in mind. First, end users do not need code-level explanations and developers can track down system processing in more efficient ways, such as by using the DEKADE environment or looking at the code itself. Second, in the big picture of developing reasoning algorithms for cognitive systems, the additional work of decorating them with select kinds of explanation-oriented metadata is so minimal as to hardly be worth mentioning. So, creating a generic code-explanation functionality is currently not on agenda.

9   Knowledge Acquisition

As should already be clear, for LEIAs to operate with understanding across many domains, they need knowledge—and a lot of it. By *knowledge*, we mean data that has been translated into the unambiguous, ontologically grounded metalanguage of the LEIA's knowledge bases. Previous chapters have shown how agents can acquire knowledge through learning. This chapter describes manual and semiautomatic methods of knowledge acquisition within the DEKADE infrastructure. These methods facilitate the acquisition of the bootstrapping knowledge that is needed to enable LEIAs to learn automatically with good results. The focus is on efficient methodologies that take the complexity of language and the world seriously.

Before we begin, a common misconception must be addressed. The need for manual work on knowledge in AI is not unique to knowledge-based systems. Data-driven systems also rely heavily on manual labor in the form of annotating and cleaning training datasets, tweaking parameters, and even recording application-specific responses, such as answers to typical questions. According to *The Economist*, as of January 2020, data preparation still claimed over three quarters of the time allocated to machine learning projects.[1] And according to Stasha (2021), Alexa Smart Assistant and Echo products perform so well because thousands of employees are working on optimizing these specific products. Similarly impressive human workforce outlays were needed to prepare IBM's *Watson* to play *Jeopardy!* However, despite Watson's spectacular win, it could not handle the ambitious follow-up applications it was set to (Lohr, 2021). In short, if one chooses to think of knowledge as a *problem*, then data-driven approaches to AI share the problem. However, since LEIAs are on a different path toward artificial intelligence than the data-driven majority, it stands to reason that the focus of manual labor is different as well.

This chapter, like the book overall, focuses on work that is either implemented in a computer program or well-specified by algorithms. This material does not exhaust all needs of LEIAs. As discussed in section 7.3, collaboration with colleagues working on modeling a variety of perception and action modalities is needed to enhance the knowledge substrate of LEIAs with multimodal, ontologically grounded descriptions that will enable agents to interpret sensory inputs in a way analogous to their current interpretation of language inputs. This knowledge will include such things as static images, video clips, and links to programs that operationalize concept detection. The LEIA knowledge infrastructure is prepared to receive such knowledge, since it is similar in kind to the procedural knowledge that supports language understanding and reasoning.

9.1 Introduction

The need for extensive high-quality knowledge and the perceived impracticality of amassing it—which has been referred to by some as the *knowledge bottleneck*—contributed to the demise of so-called *good old-fashioned AI* in the 1990s. It also fueled the paradigmatic turn toward data-driven methods, which were already gaining momentum at that time due to the spectacular technological advances in data availability, processing speed, and storage capacity. However, failing to address the knowledge problem in the spirit of cognitive modeling—by building ontologies, lexicons, explainable rule bases, and so on—has left a big hole in the AI landscape. Data-driven AI has, by necessity, avoided applications that require anything beyond analogical reasoning.

Whereas data-driven AI operates over uninterpreted big data, developers of cognitive systems typically assume that high-quality interpreted knowledge is available to support agent reasoning, and that cognitive agents will somehow translate perceived data into interpreted knowledge. However, as discussed in Section 2.8.3, few cognitive systems developers are working toward actually fulfilling these requirements, concentrating instead on general theories, specific engineering issues, or small-domain applications.

The problem with focusing on small-domain applications is that they obscure the need to account for many kinds of phenomena that are essential for scaling up. Take as an example polysemy—the fact that most words have multiple meanings, only one of which fits any given context. A typical cognitive system will include only one of the word's meanings in the agent's lexicon, as if the ambiguity problem didn't exist. Such simplifications mean that each time a system's coverage is extended beyond the original example set, the original solutions need to be thrown away. After all, a system that knows nothing about polysemy cannot spontaneously carry out disambiguation if placed in a context where words suddenly have more than one meaning. Forcing people to use controlled languages is also not viable, though the field has amply explored this option.[2] It is more efficient—in fact, imperative—to address real-world challenges more holistically from the outset, using the kinds of explanatory, evolving microtheories described in earlier chapters.

Much has changed since the early days of AI, when manual acquisition of knowledge was tried and found to be slow and cumbersome. In those days, *everything* about computing was slow and cumbersome. Now it's a completely different world with respect to processing speed, storage space, user interfaces, tools for building interfaces, and the availability of large corpora and online knowledge bases. Moreover, the newly available large language models can be used to configure various kinds of support tools. All of this fundamentally changes the prospects of manual and semiautomatic knowledge acquisition, which is a necessary complement to the ideal, but not imminent, state of affairs in which agents can learn everything automatically.

On the human front, it is becoming ever clearer what and how much to expect of knowledge workers. Typically, knowledge workers are not productive for the entire duration of a standard eight-hour workday; they are productive for only about three to six hours a day, depending on the task (Hakes, 2021). In addition, they cannot concentrate when they are bored. A stark example of the consequences of boredom-induced inattention is the fatal accident caused by an Uber autonomous vehicle, whose human operator was supposed to be vigilant enough to prevent accidents despite having nothing to do almost all the time (Smiley, 2022). Our own, more mundane experience in developing knowledge-based systems has shown that it is difficult to motivate workers to carry out knowledge acquisition unless it is divided up into small, precisely defined tasks punctuated by frequent, satisfying milestones.

It is also important to have an efficient, pleasant knowledge-engineering environment. All knowledge bases need to be easily accessible, viewable, and editable, either as a freestanding task or in conjunction with assessing agent operation. In fact, folding knowledge acquisition into a workflow that involves LEIA processing is exactly the kind of goal-oriented methodology that can keep knowledge workers engaged (see section 9.4).

This chapter discusses knowledge acquisition from three perspectives: the ontology, the lexicon, and a workflow that interleaves knowledge acquisition with agent functioning. These dovetail with the agent's learning through dialog and reading, which were described in previous chapters.

9.2 Acquiring Ontology

Ontology development involves cognitive modeling, since a LEIA's ontology must capture how people understand and reason about the world. No LEIAs are expected to be omniscient. Most of them will need highly specialized knowledge only about a particular domain, or none at all. But they all need general knowledge. For example, all dialog agents have to know the inventory of speech acts (asking a question, issuing a command, proposing a plan, and so on), and they need to know the dozens of ways of expressing each one in language. So, knowledge of this type is a high priority with a high return across applications. By contrast, many utterances and situations cannot be fully understood without domain-specific knowledge. For example, the sentence "Golfers have always walked in competitive tournaments"(COCA) implies that they don't ride in golf carts—something that might not be obvious to all readers, even those living in societies where golf is played. Specific knowledge like this is recorded in scripts.

Scripts can reflect knowledge in any domain—what happens at a doctor's appointment, how to make pizza, how to prune an apple tree; and they can be at any level of granularity—from a basic sequence of events to the level of detail needed to generate interactive computer simulations. For general domains, knowledge engineers can double as domain experts, whereas for specialized domains, they must consult outside experts. In both cases, text sources can be useful for reference.

Ontological scripts can require expressive means beyond the simple property-facet-value descriptions of basic ontological frames. Taking examples from the domain of doctors' appointments, scripts require:

- **The coreferencing of arguments.** In a given appointment, the same instance of DOCTOR will carry out many actions, such as asking questions, answering questions, and recommending interventions; and the same instance of PATIENT will carry out many actions, such as answering questions, asking questions, and deciding about interventions.

- **Loops.** There can be many instances of event sequences, such as ask/answer a question and propose/discuss an intervention.

- **Variations in ordering.** A doctor can get vital signs before or after the patient interview and can provide lifestyle recommendations before or after discussing medical interventions.

- **Optional components.** A doctor may or may not engage in small talk and may or may not recommend tests or interventions.

- **Time management.** For simulation-oriented scripts, the script must include information about what happens when, how fast, and for how long.

Whereas it should be possible for agents to automatically learn some aspects of some kinds of scripts from texts, full automation is unlikely to ever be the full answer to the problem. There are several reasons why.

1. Books and other texts intended for people do not describe how everyday life works, they provide happenstance snippets. And people do not learn how everyday life works from books, they *live*.

2. As concerns more specialized knowledge, aspects of scripts are recorded in technical manuals and textbooks, but the quality, depth, and comprehensiveness of the descriptions varies dramatically. To see an example, pull out the manual for your car or some appliance and look at the troubleshooting instructions. Do they make perfect sense—or any sense at all? It is difficult to formulate procedural knowledge because it needs to address the needs of readers with different amounts of background knowledge. Moreover, manuals intended to be used by people do not include the kinds of details that agent systems need in order to learn to operate optimally, such as all of the important subevents of scripts, constraints on their players and props, indications of required versus optional components and steps, reasons why things should be done in a particular order, allowable variations in the ordering of subevents, event preconditions and effects, and so on.

3. Scripts need to cover not only what people *do* but also *how they think about things*. This includes their abstract understanding of how life

works, how to make decisions in complex contexts, how to teach the next generation their hard-won knowledge, and so on. Although experts clearly operate with such models, they are implicit and must be made explicit through the process of knowledge engineering.[3]

To sum up, texts practically never contain anywhere near the complete script-based information that agents need to operate intelligently, even if agents could analyze all of the world's texts automatically and with perfect precision. So, manually acquiring ontology, with as much automatic support as is practicable, is a necessity.

Figure 9.1 shows some workflow options for ontology acquisition, which we will describe in turn. Lexicon acquisition typically piggybacks on ontology acquisition so that the agent can talk about the new concepts it has learned.

Knowledge engineers model general domains   For knowledge in the general domain, knowledge engineers can double as subject matter experts. Various methodologies can be followed, but in all cases the knowledge engineer must have a specific, application-oriented goal in mind. Without such a goal,

1. it would be impossible to decide what to work on: negotiation strategies, how to play baseball, or the food preferences of the world's animals;

2. it would be impossible to know the necessary and sufficient amount of detail; and

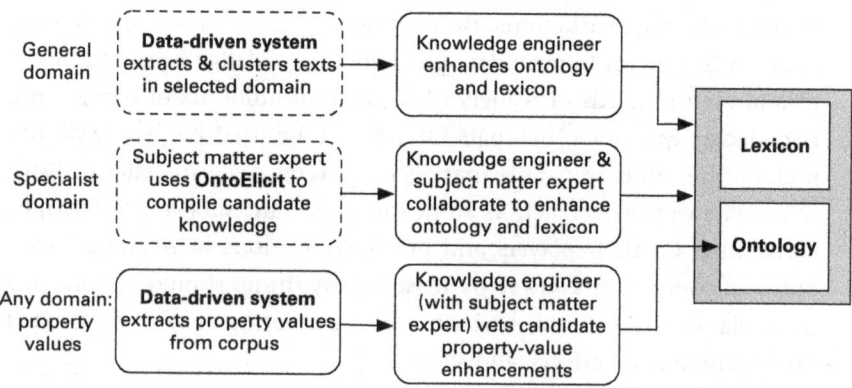

Figure 9.1
Sample knowledge-acquisition strategies that focus on ontology. Boldface indicates automatic systems, and dotted lines indicate that the modules are optional.

3. it would be impossible to validate whether the information was actually useful—that is, whether it was of the type and grain-size needed to support automatic reasoning.

Once they have identified an ontology-acquisition objective, knowledge engineers can consult a variety of knowledge sources, compiled either manually or with the help of a data-driven system that extracts and organizes texts about the given topic.

Adding new ontological concepts can necessitate adding new lexicon entries or modifying existing ones. For example, at a given time the ontology might contain only the generic concept for DOG, with the words referring to different breeds of dogs being listed as hyponyms of *dog-n1*. If an acquirer decides to expand DOG into a subtree of breeds in order to describe each one more precisely, then the words for each breed need to be removed from the hyponyms field of *dog-n1* and be promoted to their own lexical senses: *poodle-n1* mapping to POODLE, *dachshund-n1* mapping to DACHSHUND, and so on.

Knowledge engineers collaborate with subject matter experts   To model specialist domains, knowledge engineers need to collaborate with subject matter experts. Consider again the domain of clinical medicine (see section 8.8). The selection of properties to be included in a disease model is guided by practical considerations. Properties are included if they can be measured by tests, if they can be affected by medications or treatments, and/or if they are necessary components of a physician's mental model of the disease. In addition to using directly measurable properties, models also include abstract properties. For example, when the property PRECLINICAL-IRRITATION-PERCENTAGE is used in scripts describing esophageal diseases, it captures how irritated a person's esophagus is before the person starts to experience symptoms. Preclinical disease states are not subject to measurement by tests because people do not go to the doctor before they have symptoms. However, physicians know that each disease process has a preclinical stage, which must be accounted for in an end-to-end, simulation-oriented model. Coming up with useful, appropriate abstract properties reflects one of the creative aspects of computational modeling. The abstract features used for cognitive modeling are similar to the intermediate (non-leaf) categories in ontologies. Although regular people might not think of WHEELED-LAND-VEHICLE as a category, this can still be a useful node in an ontology.

Once an approach to modeling a disease has been devised and all requisite details have been elicited from experts, the disease-related events and their participants are encoded in ontologically grounded scripts written in the metalanguage of the LEIA's ontology.

Because knowledge engineering is expensive, it is well worth developing tools and automated support for the process. We developed the prototype for a tool called OntoElicit, which helps subject matter experts to record key building blocks of models before interacting with knowledge engineers. This tool encapsulates a theory and methodology of knowledge elicitation developed during two quite different projects: the Maryland Virtual Patient system and the Boas system for eliciting knowledge about low-density languages in service of machine translation (McShane et al., 2002). We use examples from the former in this thumbnail overview.

In OntoElicit, subject matter experts are led through a sequence of interactions with the system in order to complete the following tasks:

- Divide each disease into any number of conceptual stages correlating with important events, findings, symptoms, or the divergence of disease paths across patients.
- Indicate the typical duration of each stage as a range with a default value.
- List the relevant physiological and symptom-related properties, along with their typical value ranges and default values during each stage.
- List the tests that might be performed and the clinical guidelines for ordering them.
- For each test, if it is carried out at each conceptual stage, list the expected raw results and specialists' interpretations, with the latter including pertinent negatives, diagnoses, "suggestive of [disease]" statements, and so on.
- List interventions that might be performed, the clinical guidelines for ordering them, how property values are affected by the intervention if it is carried out at each conceptual stage, the possible outcomes of the intervention, possible side effects, and, if known, the percentage of the population expected to have each outcome and side effect.

The result of working with OntoElicit is the skeleton of what will become a disease model like the model of achalasia shown in tables 8.8 and 8.9.

Some kinds of agent capabilities are easier to model than others. For example, it is straightforward to prepare a tutoring agent to check if the preconditions for a move have been met, but it is more difficult to model how to select among multiple moves, all of whose preconditions have been met. In the case of clinical medicine, models of decision-making must incorporate the possibly diverging preferences of a variety of stakeholders (e.g., the patient, the physician, the insurance company), differing cost-benefit analyses for different options, and the potential need for decision-making under uncertainty since it is not unusual for some key information to be unavailable at the time a decision must be made. In order to manage this complexity, we have experimented with the use of Bayesian networks. The idea was to establish priors by asking subject matter experts to assess, for different combinations of property values, the "goodness" of different available decisions. Our experience suggests that this method of knowledge acquisition and associated reasoning merits further exploration.

Knowledge engineers use a data-driven system to suggest property values An insufficiency in the LEIA's current ontology is that some property values that should be locally specified for a concept are not; instead, an overly generic value is inherited from the parent. For example, the weight range for adult dogs (DOG) is 2–200 pounds, but the weight ranges of chihuahuas (CHIHUAHUA) and mastiffs (MASTIFF), which are children of DOG, are much narrower. Knowledge engineers can improve concept descriptions with the help of various types of automation. For example, they can use the LEIA's automatic property-learning mechanism, described in sections 7.1.3 and 7.2.2, to suggest property values, or they can use large language models (LLMs) by feeding them appropriate prompts. The main difference between the methods is that the LEIA's automatic property-learning mechanism includes a trace back to the source material, whereas LLM-based responses do not; and, in fact, the latter cannot be fully trusted. This means that LLMs are best used in cases where the acquirer has a notion of the right answer and is asking the LLM for a reminder.

9.3 Acquiring Lexicon

Another way to approach knowledge acquisition is by focusing on the lexicon and then supplementing the ontology as needed. There are many possible workflows, including those shown in figure 9.2.

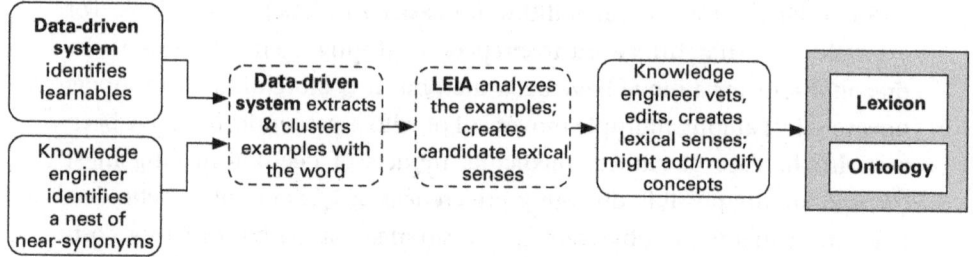

Figure 9.2
Sample knowledge-acquisition strategies that focus on lexicon. Boldface indicates automatic systems, and dotted lines indicate that the modules are optional.

The first question is how to identify which words and expressions to learn. This can be done either using a data-driven tool or by knowledge engineers working with linguistic resources such as thesauri and WordNet.

Using data-driven tools to identify learnable words and expressions The principles of the data-driven approach to learning were explained in section 7.1.3. Here we focus on matters of content.

In language, frequency of occurrence matters—a truism being explored in theoretical, computational, and corpus-based linguistics. For example, in the theoretical paradigm called the *usage-based model*, "language is seen as a probabilistic system of emergent structures and fluid constraints that are grounded in the language user's experience with concrete words and utterances" (Diessel, 2016). In other words, in human language processing, there is no stark boundary between abstract syntactic constructions and the words that can populate them. Instead, linguistic constructions are most appropriately defined at multiple levels of abstraction, including using particular words in particular syntactic structures.

For agent modeling, the following aspects of linguistic frequency are particularly important.

1. Agents need to become competent language users in the general domain as quickly as possible so that they can turn to automatically learning about specialized domains. For this, they first need to accumulate a large store of frequent expressions paired with their meanings. For example, people often make a request by saying, *I'd appreciate it if you'd VP*, so agents need to recognize this as an expression and know that it maps to

the concept REQUEST-ACTION. The corpus-attested frequency of such multi-word expressions can help to prioritize knowledge acquisition.

2. It is both useful and theoretically motivated to record the meanings of at least some very frequent multiword expressions even if they are semantically compositional. For example, *I'm hungry* occurs 1,794 times in the COCA corpus. Enabling agents to directly access language-to-meaning couplings for multiword expressions results in high-confidence analyses, enhances processing efficiency, and models our understanding of human memory and information access.

3. Identifying high-frequency constructions that include particular words can inform the learning of more abstract constructions. For instance, if a particular complete sentence occurs multiple times in a corpus, it is a candidate for being listed in the lexicon. If the LEIA uses the COCA corpus as a search space, it will find that the following full sentences, among many others, are attested multiple times:

> Dinner is served.
> Breakfast is served.
> Lunch is served.
> Tea is served.
> Justice is served.

Having extracted the full set of frequent sentences in a corpus, the agent can then cluster them and determine whether minimal pairs differ in an ontologically significant way. In the examples above, four of the five sentences involve a MEAL (*tea* can refer to a small afternoon meal), so the agent can hypothesize that $Subj_{MEAL}$ *is served* is a construction.[4] However, the agent has no way of knowing that this construction does not simply assert that a meal has been served. Instead, it is an invitation to come and eat, so its semantic description should be headed by INVITE. This is a good example of why people need to remain in the loop of knowledge acquisition.

4. It would be ill-advised to indiscriminately record fully compositional multiword expressions that are only moderately frequent, such as *(someone) had a burger*. Not only would this likely not align with people's lexical knowledge, it would also make the lexicon unnecessarily large—in a similar way as explicitly listing the passive voice of all verbal senses (cf. section 4.2.2).

5. In order for automatic processing to actually help, rather than hinder, knowledge engineering, knowledge engineers and system engineers need to work together to identify useful search strategies. Continuing with the case of multiword expressions, some rule-out conditions are clear. For example, pronoun-rich collocations like *He did it* cannot be associated with a static semantic interpretation, so there is no benefit to recording it as a multiword expression.

6. Further investigation is needed to determine in which ways frequently met-with sequences of words can vary while still having high potential for being multiword expressions whose meaning is worth recording. For example, can the words vary in morphological features? Can they be freely modified? Can they occur within larger sentences? Can any of the slots in the candidate expression be filled by a variable? If the search criteria are too strict, they will miss useful candidate expressions; if they are too loose, they will overwhelm the human who must evaluate the hypotheses.

7. There are expressions, both single-word and multi-word, that are extremely common and have a privileged status in a given type of context. For example, when customers in a restaurant or coffee shop say "Large latte," they are placing an order. Similarly, when surgeons in an operating room say "Scalpel," they are asking to be handed a scalpel. It would be useful for an automatic system to identify frequent utterances like these to remind knowledge engineers that they must be covered. Once such utterances are identified, there are several options for preparing LEIAs to correctly interpret them. On the one hand, a lexical sense can be added that asserts the given form-to-meaning correlation, but it must be appended with a meaning procedure that ensures that the context is appropriate. On the other hand, the ontological script for the given domain, such as SURGERY, can include the knowledge that when a surgeon, during a surgical procedure, names a tool, it is a request to be given that tool.

Frequency-driven knowledge acquisition is wide open territory for exploring how data-driven methods with various kinds of human guidance can speed up the acquisition of language expressions that will help LEIAs achieve basic language competency that is useful across domains and applications.

Knowledge engineers identify learnable words and expressions Human-oriented linguistic resources—grammars, thesauri, classifications—can be useful for jogging knowledge engineers' memories and helping them to organize acquisition efficiently. For example, Levin (1993) presents a classification of English verbs according to their syntactic behavior, driven by the hypothesis that verbs that are similar in syntactic behavior have semantic affinity. For example, Levin's *grow* verbs—which include *grow*, *develop*, *evolve*, *hatch*, and *mature*—are similar in that they permit an alternation between *into* and *from* (9.1) as well as a causative alternation (9.2).

(9.1) a. That acorn will grow into an oak tree.
 b. An oak tree will grow from that acorn. (Levin, 1993, p. 174, #395)

(9.2) a. The gardener grew that acorn into an oak tree.
 b. The gardener grew an oak tree from that acorn. (Levin, 1993, p. 174, #397)

Levin's verb classification can help to speed up the acquisition of verbs that have similar syntactic behavior and, often, map to the same or relatively proximate concepts.

Another example of a useful classification involves paraphrases, for which various classifications have been proposed.[5] LEIAs handle many classes of paraphrase as a matter of course: lexical synonyms, different forms of referring expressions, full and elliptical utterances, active and passive alternations, and so on. But additional phenomena must also be covered by a LEIA's knowledge bases and reasoners.

- Paraphrases can show alternations between events and social roles: *Stuart teaches our kids chemistry* versus *Stuart is our kids' chemistry teacher.*

- They can express an event or its converse: *Stuart called up Beth* versus *Beth got a call from Stuart.*

- They can express something directly or as a double negation: *Stuart wants to go* versus *Stuart doesn't not want to go.*

- They can use direct quotes or narrative: *Stuart said, sure he'd come* versus *Stuart said, "Sure, I'll come."*

- They can use light verb constructions or semantically specific verbs: *Stuart did the dishes* versus *Stuart washed the dishes.*

- They can use metonymy or a direct reference: *The red hat just smiled at Stuart* versus *The girl with the red hat just smiled at Stuart.*

LEIAs handle some of these using lexical constructions (e.g., *X called Y* and *Y got a call from X*), and others using reasoning processes (e.g., metonymy resolution). But knowledge engineers have to remember that such phenomena can occur—and that's where classifications come in handy.

Another way to approach lexicon acquisition is for people to identify, using a thesaurus or WordNet, nests of near-synonyms that are worth acquiring and then acquire them either manually or with various kinds of automatic support.[6] Table 9.1 shows some examples of near-synonyms that should map to the listed concepts—naturally, within fully specified lexical senses that include all of the necessary syntactic and semantic dependencies.

In looking at this list, one might think that compiling lists of near-synonyms should be automatable. However, as explained in section 3.1, a lot of entities listed in thesauri are not even near-synonyms. Consider some examples:

- One thesaurus lists all of the following as synonyms of *help* but they are better treated in other ways in the LEIA's knowledge bases:

  ○ *buck up* and *root for* should map to ENCOURAGE;

  ○ *stand by* and *stick up for* should map to DEFEND; and

  ○ *take under one's wing* and *open doors* are so specific that they need to be described using multiple concepts linked by properties.

- The WordNet synset (synonym set) for *scream* includes the following:

  ○ useful synonyms: *shout, shout out, yell, holler*

  ○ detrimental synonyms because they have other main meanings or are too rare in this meaning: *cry, call, hollo, squall*

  ○ useful direct troponyms: *whoop, shriek, screech, howl*

Table 9.1
Near-synonyms that map to a known concept.

HELP	ALLOCATE	WALK
do a favor	divvy up	hoof it
do a service	dole out	wend one's way
lend a hand	pass out	go on foot
COMPLAIN	ASSAULT	HIDE (oneself)
kick up a fuss	slap around	go into hiding
make a fuss	let have it	go underground
sound off	work over	lie low

○ direct troponyms that could be added to the lexicon but will be of little use because they are so rare; and, if added, they must be flagged as rare so that they will not be used in generation: *ululate, yawp, yaup*

○ detrimental direct troponyms, which are too rare or too different to be acquired at all: *screak, skread, skreigh, halloo, pipe up, pipe*

So, if data-driven tools are used specifically to identify near-synonyms—rather than, for example, to identify frequent words or multiword expressions in a corpus—their results need to be inspected by a knowledge engineer. At the time of writing, we are experimenting with using LLMs to suggest synonym-based enhancements to the lexicon.

Processing candidate additions to the lexicon Once words and expressions have been selected for acquisition, the process can unfold in various ways (cf. figure 9.2). Optionally, a data-driven system can be used to extract and cluster examples containing the word or expression. This is useful to jog acquirers' memories about meanings and usages that are not the first to come to mind. Next—also, optionally—those examples can be sent through a LEIA's language understanding and learning processes, resulting in candidate senses for the new words and expressions. Finally, knowledge engineers create—or review and edit automatically created—lexical senses, which might involve adding or modifying ontological concepts as well.

Although it might seem like automation should always prove useful, the fact is that machine-generated lists and clusters do not always speed up humans' work, as became clear in the early days of machine-assisted translation. Fully manual approaches can actually be faster and/or less frustrating, depending on both the quality of the automatic results and the preferences of individual workers.

For acquiring LEIA-style lexical senses, we expect automation to be useful mostly with respect to syntax—for example, selecting a transitive verb template based on corpus examples. Acquirers will still be responsible for vetting the semantic mapping and adding additional property values if needed, since ontological concepts are, by design, more coarse-grained than the meanings of many of the words of any language. For example, POLITENESS and FORMALITY are features that primarily apply to language, not ontology; so, values for these features are recorded in lexicon entries for the corresponding words and expressions. For example, *I would really appreciate it if you would VP* is a REQUEST-ACTION with the feature values "FORMALITY .7" and "POLITENESS 1." As with any abstract scalar properties, assigning values to particular lexemes is aimed at supporting useful reasoning, with no claim that the values reflect any precise or provable reality. The goal is for LEIAs to be as sophisticated as possible while still being developed on a fast time scale to offer near-term utility.

Some of the semantic features that distinguish near-synonyms involve the core meaning, rather than the style, of the message. For example, *rush off* can be described in any of the following ways:

– EXIT (URGENCY .8)

– EXIT (VELOCITY .8)

– EXIT (URGENCY .8) (VELOCITY .8)

The reason for the options is that this expression can imply physical speed, urgency, or both. Additional examples of an appropriate grain-size of description for the LEIA's lexicon are as follows:

– do a favor: HELP (FORMALITY .4)

– do a service: HELP (FORMALITY .8)

– lend a hand: HELP (FORMALITY .2)

– kick up a fuss: COMPLAIN (FORMALITY .1)

– sound off: COMPLAIN (FORMALITY .4)

– slap around: ASSAULT (INSTRUMENT PALM-OF-HAND)

– let have it: ASSAULT (FORMALITY .2)

– work over: ASSAULT (FORMALITY .2)

These examples show just the skeleton of the semantic side of entries. Each construction needs to be described using a full lexical sense of the type presented in earlier chapters.

To wrap up this section on acquiring lexicon, it is worth noting that it can be difficult even for people to describe certain kinds of abstract objects and events in a way that is really useful for machine reasoning: *privacy, capitulation, endearment*. Technically, it would be easy to have the agent create a new lexical sense and associated concept for each such notion without attempting to describe its distinguishing property values, but that would be kicking the can down the road and would run counter to the principles of content-centric computational cognitive modeling. We do not exclude the possibility of automatically learning some abstract notions; however, this will be most successful if they are related in some obvious way to a well-described existing concept. For example, if the ontology includes LOVE-EVENT, and the lexicon maps the verb *love* to it, then the agent can learn the meaning of the verb *adore* from the definition "to love intensely": that is, LOVE-EVENT (INTENSITY .9). However, expecting the agent to learn a concept like LOVE or ADORE from scratch, simply from its usage in various contexts, seems unrealistic. Thus, the important role of human acquirers in compiling bootstrapping-worthy knowledge bases.

9.4   Threading Knowledge Acquisition with System Operation

Knowledge acquisition can be threaded with system operation in various ways. We already saw in chapter 7 how LEIAs can learn lexicon and ontology while operating in various modes. And we saw in the previous subsections how people can acquire lexicon and ontology with various levels of participation by LEIAs. In addition, people can acquire knowledge while carrying out system testing and debugging. That is, they can run sentences through the agent's language understanding system, inspect the results using the DEKADE environment, and then enhance the knowledge bases as needed to result in a correct analysis. This enhancement can invoke various levels of automatic processing by the LEIA.

There are four reasons to thread knowledge acquisition with system operation.

1. All acquired knowledge needs to actually serve processing, and running sentences while acquiring resources is a good way of making sure that it does. For example, a word like *respectively* cannot be described using ontological concepts; it requires procedural semantic analysis provided by a custom program. If an acquirer tried to add this word to the lexicon without both specifying a meaning procedure and ensuring that it was implemented properly, then the sense could not be used by the system. Similarly, if acquirers fail to provide sufficient information in the lexicon to permit different senses of words to be disambiguated, then they are setting the agent up to face residual ambiguity every time it encounters the given word.[7]

2. Knowledge engineering is mentally tough: it is open-ended, it imposes a heavy cognitive load, and it does not offer any inherent milestones akin to a programmer's opportunity to run a program and watch it work.

Orienting acquisition around making a given input work correctly offers frequent and concrete milestones.

3. LEIAs can automatically generate certain aspects of candidate knowledge, which can speed up knowledge acquisition. For example, they can posit templates for new lexical senses that match the syntactic use of the word or phrase that is attested in the input.

4. When knowledge engineering is threaded with system operation, a side effect is creating a repository of correctly analyzed texts, that is, pairs of sentences and their correct TMRs. Such a repository, when viewed as a component of the agent's episodic memory, facilitates language processing through reasoning by analogy. This can be implemented using knowledge-based methods or machine learning. Using knowledge-based methods, the agent can consult previous correct TMRs for guidance about how to handle difficult analysis decisions in a new input. For example, many expressions have both literal and metaphorical meanings. If the agent's stored TMRs overwhelmingly prefer one over the other—given that both options are available in the given context—then that is a vote for the choosing the more frequently attested meaning. To give a concrete example, if most stored analyses of sentences of the form *X is gonna kill Y!* refer to being angry at, not killing, someone, then the angry interpretation will be the default hypothesis for future inputs of this form. However, "of this form" can be tricky to automatically compute. In this example, it is important that the verb be in the future tense because the metaphorical meaning is rarely if ever used in the past tense. The other option for implementing reasoning by analogy uses machine learning, which requires creating a large enough repository of sentence-to-TMR mappings to serve as training material.

Figure 9.3 illustrates a human-inclusive workflow that threads knowledge acquisition with system testing and enhancement. This workflow breaks the process of knowledge acquisition into small chunks with a milestone at the end of each one.

The knowledge engineer selects sentences to analyze and works on them until the analysis is correct, resulting in what is called a *golden* or *gold-standard* TMR that can be stored to episodic memory.[8] The workflow can involve lexical acquisition, ontology acquisition, and/or the improvement of analysis algorithms themselves (jointly with a software engineer).

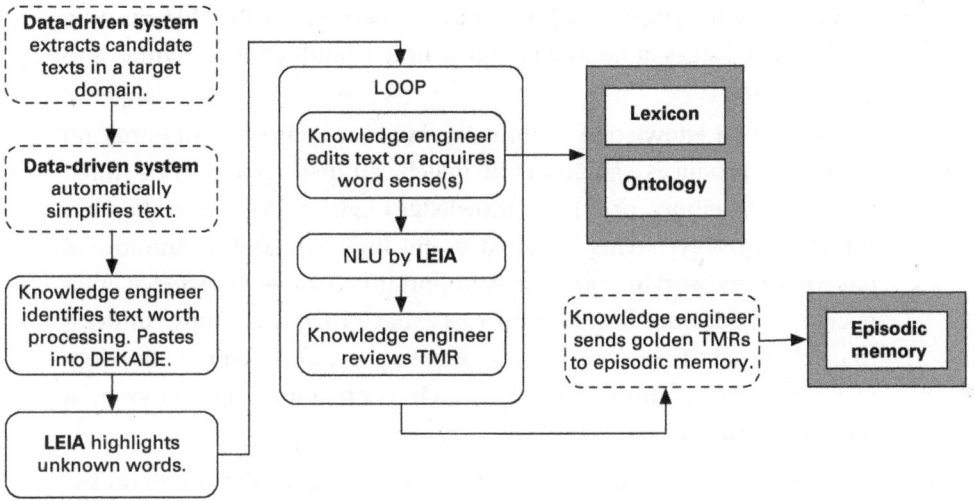

Figure 9.3
A sample human-inclusive knowledge-acquisition strategy that incorporates language processing by LEIAs. Boldface indicates automatic systems, and dotted lines indicate that the modules are optional.

Essential to this methodology is giving knowledge engineers full freedom to decide what to work on, what to postpone, and which kinds of automation to use. We will work through figure 9.3.

- **Identify a text to work on** either automatically or manually. A potential benefit of automatic extraction is that similar texts can be clustered, offering better coverage of both content and linguistic expressions at one go.

- **Simplify the text if needed**, automatically and/or manually. As Steven Pinker has pointed out, much of academic writing stinks (his term; Pinker, 2014). He attributes this, for the most part, not to the ill will of scholars or a desire to obfuscate but, rather, to the fact that writing well is hard. Moreover, as long as bad writing continues to be published, and those publications continue to advance people's careers, it must be accepted as the norm (Albert, 2004). A long history of work on automatic text simplification has resulted in potentially useful tools,[9] and our recent experimentation using LLMs for this purpose shows promise. In addition, manual simplification is actually quite fast and simple. In figure 9.3, the option of manual simplification is folded into the task of identifying text worth processing.

- **Identify unknown words.** This is done automatically by the LEIA running the first two stages of natural language understanding: Basic Syntax and OntoSyntax.

- **Loop through knowledge acquisition and text analysis** until either (a) the system produces a golden TMR, which can then optionally be stored to episodic memory, or (b) the knowledge engineer decides to abandon the text, possibly having acquired useful linguistic and/or ontological knowledge by working on it. It is important to allow knowledge engineers to discontinue working on texts that prove to be more difficult than expected—for example, texts whose processing requires development work by a software engineer, such as programming a procedural semantic routine.

Minor edits to a text can dramatically improve a LEIA's ability to understand it. The following is how a subject matter expert, gastroenterologist George Fantry, explained, in a personal correspondence, what is measured by esophageal motility tests:

> Key measurements are Distal Latency (DL) and Distal Contractile Integral (DCI). DL is the time interval in seconds from UES relaxation to where propagation of peristalsis slows (measured in seconds). DCI is a measure of the vigor of the contraction, measured as Amplitude×duration×length (mmHg-s-cm) of the distal esophageal contraction (previously utilized only amplitude).

Thanks to past collaboration with Dr. Fantry, we fully understood this explanation, but LEIAs will face numerous challenges in interpreting this passage. To start, it shows four different uses of parentheses, whose meanings must be made manifest in the TMR. Parentheses can:

- introduce an abbreviation: (DL)
- explicitly provide a measuring unit: (measured in seconds)
- implicitly provide a measuring unit: (mmHg-s-cm)
- signal an aside: (previously utilized only amplitude).

Other complexities include:

- Ellipsis: "Key measurements [*of esophageal motility tests*]" and "([*the measurement of DCI*] previously utilized only amplitude)."
- An informal turn of phrase: "from UES relaxation *to where* propagation. . . ."
- Instances of non-coreferential *the* that must be interpreted appropriately: *the time interval, the vigor, the distal esophageal contraction.*[10]
- The mathematical use of *x*.

A LEIA would be better able to derive the meaning if the input were simplified. Table 9.2 shows one such possibility.

This rewrite preserves most of the content of the original text but using a simpler writing style. We must emphasize that we are not absolving LEIAs of the need to process difficult texts. In fact, we spend a lot of time preparing them to deal with inputs that include phenomena that are outside of their current capabilities.[11] However, building agent systems requires sober practicality: it doesn't make sense to ask a LEIA to semantically analyze seriously challenging texts.

The above text simplification methodology is supported by interface functionalities in the DEKADE development environment. Additionally, as mentioned earlier, text simplification tools can be used to at least partially automate this process.

We conclude this chapter on manual and semiautomatic acquisition by reiterating the main point: if one argues for the necessity of developing explainable cognitive systems, one must have a realistic plan for endowing agents with knowledge that is of a size and complexity that makes them useful collaborators. Our plan comprises two parts. Knowledge engineers, supported by well-designed automation, work on compiling lexical and ontological knowledge that serves as a foundation. At the same time, LEIAs are being designed to learn automatically, bootstrapping from that foundation and adding knowledge at the fringes of what they know. This approach is no more labor intensive than the data-driven approaches that dominate mainstream AI.

Table 9.2
An example of text simplification.

Original Text	Simplified Version
Key measurements are Distal Latency (DL) and Distal Contractile Integral (DCI)	Esophageal motility tests measure Distal Latency and Distal Contractile Integral.
DL is the time interval in seconds from UES relaxation to where propagation of peristalsis slows (measured in seconds).	Distal Latency is the time interval between the relaxation of the upper esophageal sphincter and the slowing of peristalsis. It is measured in seconds.
DCI is a measure of the vigor of the contraction, measured as Amplitude × duration × length (mmHg-s-cm) of the distal esophageal contraction (previously utilized only amplitude)	Distal Contractile Integral measures the intensity of the contraction of the distal esophagus. The measuring unit for Distal Contractile Integral is mmHg-s-cm.

10 Disrupting the Dominant Paradigm

This book has described a human-inspired approach to modeling cognitive agents that understand their own behavior and, therefore, can explain it in normal, human terms. Any cognitive system developed using this modeling strategy is, in our parlance, a LEIA.

The fact that LEIAs are being designed to collaborate with people gives rise to certain priorities, including meaning-oriented language processing, human-oriented explanation, and lifelong learning. It also opens up certain opportunities, including distributing tasks across agents and people in a way that plays to their respective strengths and enabling agents to learn interactively, as human apprentices would. Since LEIA development is task-oriented rather than method-oriented, it accommodates a variety of metalanguage formalisms, algorithmic solutions, and means of acquiring knowledge, including taking advantage of the capabilities of large language models (LLMs).

Any specific LEIA development effort requires a host of technical and methodological decisions, and our group's status as a university laboratory naturally affects such decision-making. For example, we prioritize scientific discovery over resource acquisition, we attempt to solve problems in a manner that establishes a theme for which technologists can work out the variations, and we develop demonstration systems as deliverables for funded research.

The title of this book was selected judiciously. Achieving trustworthy AI *will* be a long game, and it would be a mistake for society to entrust certain kinds of responsibilities to systems that are not operating with intentional intelligence. Key to this statement is the modifier: *certain kinds of* responsibilities. Machine-learning-based AI systems produce useful results, including some that contribute to LEIA operation. However, in our estimation, machine learning is not the whole answer to achieving humanlike artificial intelligence. This is the nontraditional claim that renders the LEIA program of R&D disruptive.

In the popular press, there is a concerning blurring of the line between scientific reality and sci-fi fantasy. The popular imagination is naturally animated by the singularity scenario, the idea of machines becoming Terminators, and the notion that machines could become so fixated on accomplishing a task—like making paper clips in Nick Bostrom's thought experiment—that they would pursue it to the point of harming humanity.[1] Although such imaginings make for lively dinner conversation, they do not realistically predict the trajectory of AI development.

Similarly problematic is the overreliance on metaphors both within the AI community and as a tool for explaining AI to non-specialists. Metaphors can help scientists think about complex problems, but they can also become so entrenched as to inhibit creative thought. They can provide the public with insights into scientific discoveries, but they can also create misunderstandings due to the public's inability to recognize the distance between the metaphor and the reality it seeks to evoke. The professional and popular literature on AI is replete with metaphors, starting with the term *artificial intelligence* itself and proceeding to such things as *neural nets*, *machine learning*, *machine reasoning*, *computer vision*, *chat bots*, and *adversarial AI*. The metaphorical links of AI terminology to human experience inevitably contribute to unrealistic assessments by non-specialists of both the power of modern AI and the threats it engenders.

One metaphor that is particularly misleading is that ML-based AI systems learn automatically, which implies without the need for human labor. However, in reality, ML-based AI relies on a massive workforce of human data annotators. Josh Dzieza[2] paints a stark picture of the work lives of these annotators: their job is difficult, repetitive, and boring, and the workers are given no insight into how their efforts will contribute to any tangible goal. Dzieza explains, "When AI comes for your job, you may not lose it, but it might become more alien, more isolating, more tedious." Although LEIA R&D also requires human labor, it is of a completely different kind: trained knowledge engineers will enhance interpretable knowledge bases in particular ways for clearly identified purposes. This makes knowledge engineering more akin to the desirable profession of software engineering than to the grueling task of corpus tagging.

It is not only metaphors that mislead; so can anthropomorphizing. As an article in *The Economist* points out, humans have an ancient tendency to anthropomorphize, but it should be resisted when thinking about AI technologies like LLMs. "These systems, including those that seem to converse, merely take input and produce output. At their most basic level, they do nothing more than turn strings like 0010010101001010 into 1011100100100001 based on a set of instructions. Other parts of the software turn those 0s and 1s into words, giving a frightening—but false—sense that there is a ghost in the machine."[3]

How AI will play out over time depends on unknowns involving science, technology, and how society's decision-makers choose to allocate resources. Those decisions, in turn, will involve not only the decision-makers' priorities and cost-benefit analyses but also their understanding of the different technologies on offer. The latter makes it imperative for the scientific branch of the AI community to explain the different technologies plainly so that real people know what they are getting—and, more importantly, what they are putting their faith in. *The Economist* formulates this idea as follows:

> AI is too important for loose language. If entirely avoiding human-like metaphors is all but impossible, writers should offset them, early, with some suitably bloodless phrasing. "An LLM is designed to produce text that reflects patterns found in its vast training data," or some such explanation, will help readers take any later imagery with due scepticism. Humans have evolved to spot ghosts in machines. Writers should avoid ushering them into that trap. Better to lead them out of it. (*The Economist*)[4]

Since nobody can predict how AI will evolve over the next five years, not to speak of the next twenty or fifty, it would be unwise for society to place all its bets on a single theoretical or methodological approach. LEIAs offer a counterpoint to mainstream, data-driven AI, with the potential to earn an unprecedented, and critically needed, level of trust by the humans responsible for their operation.

Notes

Acknowledgments

1. Grants include *The Language of Learning*, ONR award #N00014-23-1-2060; *Natural Language Understanding by Intelligent Collaborators*, ONR award #N00014-19-1-2708; *Improving System Engineering Using Knowledge Engineering with UML*, ONR award #N00014-20-1-2051; and *Apprentice Agents*, ONR award #N00014-17-1-2218.

Chapter 2

1. See Nirenburg, McShane, and English (2023).

2. When discussing the two-system view, we refer to Kahneman's book because it offers an accessible treatment for general readers. Readers with more than a casual interest will find more technical discussions of this topic in, for example, Sun (1995), Sun and Alexandre (1997), Sun and Bookman (1994), and Sun, Slusarz, and Terry (2005).

3. See, for example, Goldwater, Jurafsky, and Manning (2010), and Rosenberg and Hasegawa-Johnson (2020).

4. For discussion, see McShane and Nirenburg (2021), section 1.4.4.1.

5. Nirenburg, English, and McShane (2021).

6. Nirenburg and Wood (2017); Nirenburg et al. (2018).

7. McShane and Nirenburg (2021), chapter 8.

8. Cognitive scientists have been studying how people visually simulate as they think (see Elder and Krishna, 2014, for a literature review). The LEIA modeling strategy neither contradicts nor is directly aided by this analysis of human cognition. With or without a mental image, LEIAs need to ground objects and events in their ontology in order to reason about them.

9. See McShane and Nirenburg (2021), section 2.8.1, for options in knowledge representation and the reasons why it is worth the substantial effort to use a language-independent metalanguage of meaning representation.

10. For more on XMRs, see English and Nirenburg (2019).

11. Upper-case semantics refers to the practice, undertaken by some researchers in formal semantics and reasoning, of asserting that strings written using a particular typeface—often, uppercase—have a particular meaning. For example, TABLE might be said to refer to a piece of furniture rather than a chart or the act of delaying the discussion of something. Upper-case semantics allows practitioners to avoid natural language challenges like ambiguity and semantic non-compositionality, and to avoid the need to actually describe the corresponding meaning in detail—which would require making it an element in a conceptual system, such as an ontology, and describing it using property values within that system.

12. This is a simplification to keep the exposition on track. The agent can actually choose to not remember some things, such as input that is outside of its interests or cannot be confidently interpreted.

13. Many aspects of the vast topic of *reasoning* have not yet made it on agenda for LEIAs. This includes, for example, the important role of formal logic in human cognition. See, for example, Bringsjord, Giancola, and Govindarajulu (2023).

14. https://www.mirror.co.uk/news/weird-news/real-life-robinson-crusoe-reveals -11976781

15. See Nirenburg and Wood (2017), Nirenburg et al. (2018), and Nirenburg, English, and McShane (2021).

16. Some practitioners claim to be working on System 2 reasoning using machine learning; for discussion, see Bengio (2019) and Dickson (2019).

17. Relevant points of comparison are Marr's (1982) three levels of understanding in information processing and Newell's (1982) classification of computer system description levels, which includes a "knowledge level." In-depth comparisons are beyond the scope of this book. See also Frigg and Hartman (2020) and Hardcastle and Hardcastle (2015). For an earlier discussion of this material, see McShane and Nirenburg (2023).

18. Our conception of a model is strategically congruent with the views of Forbus (2019, Chapter 11), though we concentrate on modeling less observable phenomena than those in the focus of Forbus's presentation. Our views on the nature of theories, models, and systems have been strongly influenced by Bailer-Jones (2009). However, for the practical purposes of our enterprise, we do not see a need to retain the same fine grain of analysis as Bailer-Jones' philosophy-of-science viewpoint.

19. To take an example from linguistics, most analyses of pronominal coreference cannot be implemented in computer systems because they require unobtainable prerequisites, such as knowing the discourse structure of the text. For discussion, see McShane and Nirenburg (2021), section 1.4.3.4.

20. See, for example, Kahneman (2011) and Gigerenzer (2008).

21. DARPA is the Defense Advanced Research Projects Agency of the United States.

22. This list is a partial inventory of relevant metaparameters. They have all been discussed in the literature:

- Simplicity has been addressed both directly (Culicover & Jackendoff, 2005) and from the opposite perspective—complexity (Newmeyer & Preston, 2014).
- Parallelism has been explored both within the language system itself (Goodall, 1987; Hobbs & Kehler, 1998) and in broader contexts, such as poetics and rhetoric (Fox, 1977; Jakobson & Vine, 1985).
- Prefabrication manifests, for example, in grammatical constructions, which are studied within the various threads of construction grammar (Hoffmann & Trousdale, 2013).
- Ontological typicality is an idea that stems back at least to Schank's work on scripts (e.g., Schank & Abelson, 1977) and "memory organization packets" (Schank, 1982).

23. In fact, there are some exceptions to the generalization that clauses in a gapping construction need to be strictly parallel syntactically, as shown by the example *One team **erred** conceptually and the other __, in practical terms*. Here, an adverb (*conceptually*) is juxtaposed with a prepositional phrase (*in practical terms*). For further discussion of the usage conditions of gapping crosslinguistically, see McShane (2005).

24. See McShane (2005; 2018), among others.

25. An example is the MUC-7 Coreference Task (Hirschman & Chinchor, 1997). For further discussion, see McShane and Nirenburg (2021, sections 1.6.8, 1.6.11, and 9.1).

26. For discussion and references, see McShane and Nirenburg (2021), chapter 9.

27. We are grateful to Dr. Michael K. Qin and the Office of Naval Research for award #N00014-20-1-2051, *Improving System Engineering Using Knowledge Engineering with UML*, which allowed us to explore the use of graphics for combined computational cognitive modeling and system engineering.

28. For example, surveys on trust and explainability include Barredo Arrieta et al. (2020), Dwivedi et al. (2022), Kaur et al. (2022), Mueller et al. (2019), and Nauta et al. (2023).

29. See Jurafsky and Martin (2023), chapters 22, 24, 26, and 27.

30. For an overview, including historical references, see Clark (2015).

31. Goyal et al. (2013) reports that work is underway to extend distributional semantics to exploit compositionality.

32. See, for example, Valmeekam et al. (2022).

33. Examples include GPT (Brown et al., 2020), LLaMa (Touvron et al., 2023), PaLM (Chowdhery et al., 2022), and BLOOM (Scao et al., 2022).

34. See, for example, Bubeck et al. (2023), Friedman (2023), Kaddour et al. (2023), Mahowald et al. (2023), and Mitchell (2023).

35. See, for example, Babic et al. (2021).

36. Machine translation systems perform poorly if they do not have access to sufficiently large parallel corpora, if the genre is informal dialog, if a text is from a domain or genre for which the system was not explicitly trained, or if the source language does not specify feature values that need to be asserted in the target language.

37. See, for example, Gunning (2017) and Mueller et al. (2019).

38. "For AI, data are harder to come by than you think," June 13, 2020, *The Economist*. https://www.economist.com/technology-quarterly/2020/06/11/for-ai-data-are-harder -to-come-by-than-you-think

39. "For AI, data are harder to come by than you think," June 13, 2020, *The Economist*. https://www.economist.com/technology-quarterly/2020/06/11/for-ai-data-are -harder-to-come-by-than-you-think

40. See, for example, Babic et al. (2021), Ehsan et al. (2022), and Liao and Varshney (2022).

41. See, for example, Liao et al. (2021).

42. Terms are used as follows: *orthotic* (Nirenburg, 2017; Wilks, 2019), *human-in-the-loop AI* (Zanzotto, 2019), and *human-AI systems* (Mueller et al., 2019).

43. Sun (2023) is a reference volume on computational cognitive science.

44. See, for example, the robotic learning application described in Scheutz et al. (2017).

45. This is true even of contributions that are expressly devoted to knowledge, such as Jacobsson et al. (2007).

46. Kotseruba and Tsotsos (2018) offer a survey. Sloman and Scheutz (2002) posit thirteen rubrics as dimensions for comparing cognitive architectures.

47. Actually, there have been past attempts to justify simplifications about language processing on theoretical grounds. One such is Christiansen and Chater's (2016) "now-or-never bottleneck" hypothesis, which they claim to be "a fundamental constraint on language." The now-or-never principle essentially states that all decisions in language processing must be local. Developers of cognitive systems have found this a convenient principle (and slogan). For example, McDonald (1983) followed this principle in the Mumble language generation system. The problem is that the claim falls apart when exposed to the actual phenomena of natural languages—as convincingly argued in published responses to Christiansen and Chater. To give just

a taste: their theory cannot accommodate nonlocal dependencies, which are a core design feature of natural languages (Levinson, 2016); it does not account for how downstream input affects the interpretation of earlier material, "which shouldn't occur if chunking greedily passes off the early information to the next level" (Mac-Donald, 2016); it does not account for many pragmatic phenomena that support the communicative function of language, such as clarification, repair, long-distance repetition, and balancing the needs of speakers with those of listeners (Levinson, 2016; Healey et al., 2016; Lewis & Frank, 2016); and their "very bottom-up characterization of chunking is inconsistent with evidence for top-down influences in perception" (MacDonald, 2016).

48. In their analytical survey of cognitive architectures, Lieto, Lebiere, and Oltramari (2018) describe knowledge-related gaps in some of the most prominent of today's cognitive architectures.

49. Developing prototypes is categorically different from non-computational descriptive work. An example of the latter is Brachman and Levesque's (2022) treatment of commonsense reasoning, in which they say, "our intent is to provide an outline only . . . in nowhere near enough detail for an engineer tasked with building a computational system along these lines" (p. 67).

Chapter 3

1. This discussion in part overlaps with McShane and Nirenburg (2021) section 1.6.7.

2. For an assessment of this program of work, see Ide and Véronis (1993).

3. For computational work involving thesauri, see Inkpen and Hirst (2003).

4. https://wordnet.princeton.edu/documentation/wnstats7wn; accessed on January 23, 2023.

5. Counts are as of June 2, 2022, using the online version available at http://wordnetweb.princeton.edu/perl/webwn.

6. The text in these and subsequent definitions is quoted directly but we present it with different formatting and punctuation than in the online version of WordNet—thus the lack of quotation marks.

7. Opinions differ about the optimal inventory of case roles for natural language processing systems. Some knowledge bases use a very large inventory of case roles. For example, O'Hara and Wiebe (2009) report that FrameNet uses over 780 case roles and provides a list of the most commonly used 25. Other resources underspecify the semantics of case roles. For example, PropBank (Palmer et al., 2005) uses numbers to label the case roles of a verb: Arg0 and Arg1 are generally understood to be the agent and theme, but the rest of the numbered arguments are not semantically specified. This approach facilitates the relatively fast annotation of large corpora,

and the resulting annotations support investigation into the nature and frequency of syntactic variations of the realization of a predicate. However, this underspecified system does not permit automatic reasoning about meaning to the degree that an explicit case role system does.

8. See the entry "Implicatures" in *The Stanford Encyclopedia of Philosophy* at https://plato.stanford.edu/entries/implicature/

9. For an in-depth treatment of ontological properties, see the "Properties" entry in *The Stanford Encyclopedia of Philosophy* at https://plato.stanford.edu/entries/properties. Note that in the original formulation of the theory of Ontological Semantics (Nirenburg & Raskin, 2004), which focused on natural language understanding outside of an agent environment, all ontological properties were treated as primitives. The need to explain them has arisen as the program of research expanded into holistic agent modeling, including learning and nonlinguistic reasoning.

10. https://www.theupsstore.com/pack-ship/moving-boxes-supplies; retrieved on August 30, 2022.

11. A sample calculation for *large*: (The range of values: 18 inches) * ("large" .8) = 14.4 inches; add this to the starting point of the scale, which is 6, resulting in 20.4 inches.

12. For further discussion, examples, and calculations, see McShane and Nirenburg (2021), section 4.1.2.

13. https://blog.dupontregistry.com/features/top-25-most-expensive-cars/, https://www.kbb.com/car-news/2022-kia-rio-keeps-price-under-17k/, accessed on June 9, 2023.

14. It would be more precise to indicate that the reference to patients should be generic; however, a full microtheory of generic references remains to be developed.

15. As the qualitative reasoning community has argued since its inception (e.g., Hayes, 1979; Forbus, 1984), in order to emulate human cognitive functioning, agents must have commonsense reasoning. As concerns properties, commonsense reasoning must (a) embrace the coexistence of qualitative and quantitative representations and (b) include heuristic decision rules that can operate over both kinds of representations. Fleshing out the interaction between qualitative and quantitative representations is especially important for language-endowed intelligent agents because meanings expressed in language—even in specialized, scientific and technological discourse—are often less precise than what heuristic decision rules in typical AI systems expect.

16. Inspiration for how to record this information comes from related work on approximations, reported in McShane and Nirenburg (2021, section 4.1.2). In different domains, approximations can carry different meanings. For example, whereas a baby who is *about 2 months old* is, at most, a week or two either side of two months,

a man who is *about 80 years old* can be several years either side of 80. Similarly, *around 8:00* allows for 10 or even 15 minutes either way, whereas *around 7:14* allows for a minute or two, max.

17. If the value of the instance is known, it is used. Otherwise, the *default* or middle of the *sem* range recorded in the ontology is used.

18. The calculations use the following measures: pencil length—7.5 inches; notebook length—11 inches; notebook width—9 inches; car length—14.7 feet; car width—5.8 feet; stop sign height—7 feet; average town size (very approximately)—300 square miles, so 17.5 miles long and 17.5 miles wide.

19. For early work on scripts, see Minsky (1975), Schank and Abelson (1977), Charniak (1972), and Fillmore (1985).

20. For an illustration of scripts acquired by agents on the fly, see Nirenburg and Wood (2017).

21. We hide the indication of facets in scripts in order to make it easier to read the structures.

22. For details, see McShane et al. (2015).

23. HAS-GENDER is a literal attribute, which is why its filler is in plain text (it is a literal, not a concept).

24. The article *the* has many functions in English; see McShane and Nirenburg (2021), section 5.4.

25. The semantic constraints on the object of *with* exclude ANIMALS, so the INSTRUMENT interpretation will not be used for the input *Fred secured the tent with Harry*, in which *Harry* is another AGENT.

26. For more on sets, see McShane and Nirenburg (2021), section 5.1.5.

27. Chapter 9 of McShane and Nirenburg (2021) describes past experiments to measure progress on natural language understanding by LEIAs. One fully expected takeaway was that a larger lexicon is needed to improve the processing of texts in the open domain.

28. See, for example, McShane, Nirenburg, and Beale (2005) and McShane and Nirenburg (2012).

29. See, for example, Nirenburg, Somers, and Wilks (2003).

30. The opticons used for past LEIA demonstration systems were specific to particular simulation and robotic environments. Work to generalize such knowledge bases is underway.

31. A similar direction of work is discussed in Krishnaswamy and Pustejovsky (2016) and Pustejovsky and Krishnaswamy (2016).

32. See, for example, Beatty (1982), Marín-Morales et al. (2020), Kahneman and Beatty (1966), Nunnally et al. (1967), and Thayer et al. (2009).

Chapter 4

1. Currently we use spaCy (Honnibal & Montani, 2017) for sentence splitting, tokenization, lemmatization, part-of-speech tagging, morphological analysis, dependency parsing, named-entity recognition, and coreference. For the coreference engine, see "End-to-end Neural Coreference Resolution in spaCy" (Oct 5, 2022, available at https://explosion.ai/blog/coref) and Lee et al. (2017). We use Benepar (Kitaev et al., 2019) for constituency parsing. However, other tools could be used as well.

2. Skimming can be implemented in various ways. For example, after Basic Syntax the agent can see if the input contains any words of interest; after OntoSyntax it can see if there are any concepts of interest; and after Basic Semantics it can see if there are any dependency structures of interest. For more examples of actionability decisions during language analysis, see McShane and Nirenburg (2021), section 2.5.

3. It is also possible to create separate lexical senses for each number of each type of conjoined entity: for example, two noun phrases conjoined by *and*, three noun phrases conjoined by *and*, and so on.

4. Prep-parts are verbal particles that have the same form as prepositions.

5. This accounts for nonstandard uses of prepositions, which can reflect misspeaking by a native speaker, an error by a nonnative speaker, or a shift in language norms. For details, see McShane and Nirenburg (2021), section 6.2.2.

6. Our microtheory does not, and will not in the foreseeable future, cover the kinds of examples that are used in theoretical paradigms to test the extremes of the human language capacity, such as *Was the doughnut given to Sally—and then accidentally dropped and eaten by her dog before being noticed by anyone—baked by your cousin?*

7. The work will necessarily include investigating how imported data-driven parsers handle inputs of increasing syntactic complexity. If the parses of complex sentences do not align with what our theory anticipates, this incompatibility will need to be handled at the system level of our theory-model-system triad (cf. section 2.2).

8. For details, see McShane and Nirenburg (2021), section 7.2.

9. For further discussion, see McShane and Nirenburg (2021), section 3.3, "Managing Combinatorial Complexity."

10. For the most part, construction semantics works over individual sentences though some transformations (e.g., for sentence-initial conjunctions) and some

Lexically Triggered Procedural Semantic routines (e.g., to resolve coreference) do operate cross-sententially.

11. For surveys of natural language generation, see Gatt and Krahmer (2018) and Santhanam and Shaikh (2019).

12. See McShane and Leon (2021) for our group's early work on language generation.

13. Ultimately, other kinds of context-sensitive referring expressions need to be added to the candidate space as well, such as *this one, the one with the ball, the other one*, and so on.

14. As with other outsourced capabilities, this software package could be replaced by others.

15. This selector was developed by LEIA lab member Sanjay Oruganti.

16. Past work includes, for example: exploration of the Incremental Algorithm (Dale, 1989; Dale & Reiter, 1995) and its extensions, such as DIST-PIA (Williams & Scheutz, 2017); analyses of particular linguistic issues, such as the need to avoid false conversational implicatures through the choice of referring expressions (Reiter, 1990); analyses of the deficiencies of proposed algorithms, including in comparison to human studies (Deemter et al., 2012); and linguistic studies in various schools of functional and discourse grammar (e.g., Keizer, 2014).

17. In fact, one reason why syntax was so long the exclusive focus of theoretical linguistics is that semantics is far more difficult to model formally.

18. For efforts to build so-called *constructicons*, see Fillmore, Lee-Goldman, and Rhodes (2012), and Lyngfelt et al. (2018).

19. See, for example, Bryant (2008), Bergen and Chang (2013), Eppe, Trott, and Feldman (2016), Eppe et al. (2016), and Feldman (2020).

Chapter 5

1. For a book-length treatment of the history of coreference studies in natural language processing, see Poesio, Stuckardt, and Versley (2016). The concluding chapter, "Challenges and Directions of Further Research," is particularly noteworthy.

2. Specifically, chapter 5 and sections 4.8, 7.7, 9.2.4, and 9.2.5.

3. See Levesque, Davis, and Morgenstern (2012).

4. Parallelism has been explored in linguistics (e.g., Hobbs & Kehler, 1998) as well as poetics and rhetoric (Fox, 1977; Jakobson & Vine, 1985).

5. In Russian and Polish, objects and function words can often be elided in coordinate structures (McShane, 2000; 2005).

6. Simplicity—particularly grammatical simplicity—has been studied in linguistics more broadly (e.g., Newmeyer & Preston, 2014)

7. Examples with the subscript "Gigaword" are from the Gigaword corpus (Graff & Cieri, 2003).

8. Examples with the subscript "COCA" are from the Corpus of Contemporary American English (Davies, 2008–).

9. For our earlier work on difficult aspects of coreference, see McShane and Babkin (2016a, b).

10. This figure is intended only to convey the gist of the ellipsis-sponsor relationship to general readers. It does not reflect a theoretical commitment to a particular tree structure, for example, with respect to binary vs. nonbinary branching.

11. Conventions used in this example set:

- The word that heads the construction, which is a non-variable anchor for the lexical sense, is in all caps.
- Synonyms of constituents are in angle brackets.
- Subscripts indicate coreference between the subject (Subj) of the main clause and the necessarily pronominal subject (Pro) of the embedded clause.
- Parentheses indicate optionality.
- "Modals" indicates one or more modal verbs.
- V^1 indicates the verb and its internal arguments and adjuncts.
- The sponsor for the ellipsis is in boldface.

12. For formal reasons, different lexical senses are needed to cover synonyms that use multiword expressions in place of *can*, such as *is able to*.

13. Certain aspects of conjunction processing are treated using syntactic transformations. For example, conjunction structures can contain any number of conjuncts, and the subjects of the latter conjuncts can be unexpressed, as in *He wanted to come but couldn't*. This chapter does not get into this level of detail.

14. An engineering note: the decision to record coreference resolution algorithms in lexical senses for function words like conjunctions and punctuation marks required only one minor modification to the LEIA's language understanding algorithm. It now needs to run multiple cycles of meaning procedures since the output of one can serve as input to another. For example, when the sense of *but* that covers "Clause1 but Clause2" identifies the sponsor for an elided VP, that sponsor must be further evaluated by the Semantic Analysis Procedures listed in the bare-modal sense that licensed the ellipsis. There are various ways to engineer the interdependence of meaning procedures. One way would put a greater burden on knowledge engineers: They would need to foresee dependencies between meaning procedures and establish the order in which to run them. Another way

would put a greater burden on processing: all procedures would need to be rerun until no new results were generated in a given cycle. Following our overarching priorities of keeping all processing as generic as possible and reducing complexity for the humans working on LEIAs, we are currently using the processing-heavy solution.

15. In an intermediate stage of the TMR, after Basic Syntax, the elided VP is understood as an underspecified event, EVENT-1, which is tagged with metadata indicating that it requires coreference resolution. The coreference procedures are run during Lexically Triggered Procedural Semantics, and it is the result of that processing that is shown here.

16. For spoken language, speech-to-text transcription needs to convey the pause, either through a punctuation mark or by a feature that can automatically be translated into a punctuation mark.

17. Further study is needed to determine the constraints on how much and what kind of text can separate the paired modals without reducing the likelihood that their VP complements are coreferential.

18. The ellipsis clause can also refer to properties not mentioned in the sponsor clause since unification still holds: Jim tried to **open the bottle** but couldn't __ without a bottle opener.

19. See McShane and Nirenburg (2021), chapter 6.

20. See McShane and Nirenburg (2021) for discussion of other kinds of referring expressions.

21. For a survey of past work on event coreference, see Lu and Ng (2018).

22. There is another sense for nonhuman animals.

23. For work on natural language understanding in Russian, see Boguslavsky (2021).

Chapter 6

1. McShane, English, and Nirenburg (2021) provide additional historical background and references, and Traum (1999; 2000) provides substantial reviews of the early literature.

2. See, for example, Traum (1999; 2000).

3. Dialog acts are so widely used in empirical NLP that over a decade ago this was already the subject of a survey article (Král & Cerisara, 2010).

4. For the Switchboard corpus, see Godfrey, Holliman, and McDaniel (1992).

5. This information about coreference is available if applicable. Patients can also describe their medical complaints without being asked.

6. See, for example, Kahneman (2011) and Bailer-Jones (2009).

7. For more on decision functions, see McShane (2014).

Chapter 7

1. The term *learning* is used in several different senses in AI. Examples of other definitions include:

1. The methods of computation collected under the rubric *machine learning*.

2. In the context of DARPA's Machine Reading Initiative, *learning by reading* refers to automatically acquiring data that the system decorates with features that enhance applications. Strassel et al. (2010) explain that the goal of the DARPA initiative was detecting select features in texts rather than comprehensive semantic analysis: "The [Machine Reading] program is structured around a roadmap of linguistic and semantic capabilities, e.g. dealing with anaphora, causal and modal language, temporal and spatial reasoning, sentiment and belief." These features are among those that were worked on in earnest as individual capabilities within mainstream natural language processing. See Forbus et al. (2007) and Barker et al. (2007) for reports on the machine reading initiative.

3. In the collocations *lifelong learning* and *continuous learning*, learning refers to the ability of neural networks to learn a new classification task without entirely losing the ability to do a previous classification task (Parisi et al., 2019).

2. See McShane and Nirenburg (2021), section 8.1.3.2

3. Our LLM-based tools were developed by Sanjay Oruganti. Past paraphrase-oriented capabilities, which were developed for applications like question answering, summarization, plagiarism detection, and authorship detection, cannot support agent learning of the kind we describe (for a literature review, see Vrbanec and Meštrović [2020]). The paraphrase corpora reported in Dolan, Brockett, and Quirk (2005) and Burrows, Potthast, and Stein (2013) are also not directly useful to LEIAs since "paraphrase" is defined very broadly.

4. Related literature includes Pantel, Ravichandran, and Hovy (2004); Nirenburg, Oates, and English (2007); English (2010); Mitchell et al. (2018); and Nirenburg, Krishnaswamy, and McShane (2023).

5. See, for example, Mitchell et al. (2018), and Mazumder, Ma, and Liu (2018).

6. Parisi et al. (2019).

7. For earlier presentations of this material, see Nirenburg and Wood (2017), Nirenburg et al. (2018), and McShane and Nirenburg (2021), section 8.3.

8. The language utterances understood by robots have been tightly constrained, with most research efforts focused on enabling robots to learn skills through demonstration (e.g., Argall et al., 2009; Zhu & Hu, 2018). The robotics community has not willfully disregarded the promise of language-endowed robots; rather, it has understandably postponed the challenge of language understanding, which, in an embodied application, must also incorporate extralinguistic context—what the robot sees, hears, knows about the domain, thinks about its interlocutor's goals, and so on.

9. For more on HTNs, see Erol, Hendler, and Nau (1994).

10. Scheutz et al. (2013) discuss methodological options for integrating robotic and cognitive architectures and propose three interfaces between them—the perceptual interface, the goal interface, and the action interface. In our work, the basic interaction between the implicit robotic operation and explicit cognitive operation is supported by interactions among the three components of the memory system of the robotic LEIA.

11. The prototype system involved simulated vision as well, but the details of visual grounding are beyond the scope of the current discussion. This scenario was first described in Nirenburg, English, and McShane (2021). The current presentation fleshes out additional details and incorporates the scenario into the generalized learning flow described in this chapter.

12. There is another lexical sense for *I'll teach you what to do if/when Clause.*

13. There are other lexical senses for other syntactic configurations, such as *You can recognize DirectObj based on NP <because of NP, based on its having NP, etc.>.* We have found it counterproductive to bunch lexical senses with significantly different syntactic configurations, even if they convey the same semantics.

14. For details, see McShane and Nirenburg (2021), section 4.1.2.

15. An important technical note is that during a LEIA's operation, new word senses that are mapped to generic concepts are not subject to scoring penalties based on selectional constraints since they would predictably fail to meet them. For example, if the unknown word *assessor* fills the AGENT slot of a SPEECH-ACT (e.g., *The assessor said . . .*) and the agent initially learns it as some kind of OBJECT, it will fail to meet the animacy requirements of the AGENT slot of SPEECH-ACT.

16. See, for example, Goldberg (2019) and Veale (2012).

17. Goldberg (2019: p. 29) discusses the example "She mooped him something", which uses the made-up verb *moop.*

18. Inspired by Goldberg's (2019: p. 37) example number 3.18, "He sneezed the bullet out of his right nostril."

19. Inspired by Goldberg's (2019: p. 37) example number 3.20, "She *kissed* him unconscious."

20. See McShane (2005), chapter 7, for discussion of generic references to humans.

21. For a more thorough explanation of why agents do not begin the analysis of every sentence using situationally informed disambiguation, see the introduction to chapter 7 in McShane and Nirenburg (2021).

22. To speed up knowledge acquisition from domain experts, intelligent systems can stand in for the knowledge engineer for the initial stages of modeling, and this semiautomatic process can result in positing new properties. See section 9.2 for a description of the OntoElicit system.

23. For details, see McShane and Nirenburg (2021), sections 4.5 and 6.3.1.

24. The reasoning functions attached to LINGUISTIC-POSSESSIVE and NOMINAL-COMPOUND-RELATION remain to be developed.

25. References include the literature and annotations associated with Rhetorical Structure Theory (https://www.sfu.ca/rst/01intro/intro.html; Das and Taboada, 2018); the corpus-annotation effort reported in Carlson, Marcu, and Okurowski (2003); and related psycholinguistic literature, such as Marchal, Scholman, and Demberg (2020).

26. For discussion, see McShane and Nirenburg (2021), section 2.8.1.

27. "AI is a lot of work," *New York Magazine*, June 20, 2023. Accessed at https://nymag.com/intelligencer/article/ai-artificial-intelligence-humans-technology-business-factory.html on June 23, 2023.

Chapter 8

1. See, for example, Malle (2004).

2. See, for example, Spaulding (2020).

3. Expanding all subtrees would have resulted in a long list of concept names that are not entirely self-explanatory.

4. Since the lexicon is a complex knowledge base, there are plenty more kinds of questions that the agent could be prepared to answer. For example, a person could want to know all of the idiomatic constructions using a particular word, the selectional constraints on the case roles in a word sense, all of the words and expressions with a particular meaning, and so on. However, enabling agents to field these kinds of questions is not a first priority for three reasons: (1) such information is too technical to be of interest to typical end users; (2) the developers who need this information can access it directly through the DEKADE environment; and (3) for many such queries, a visual representation—that is, looking at the inventory of lexical senses in an interface like DEKADE—will be more helpful than a prose response.

5. For further discussion see, for example, Kahneman (2011) and Korte (2003).

6. This is a sampling of cognitive biases. For an earlier presentation of this material with a different emphasis, see McShane and Nirenburg (2021), chapter 8.

7. The formulation of the last clause of this sentence could be smoothed, but the variable-inclusive template used to generate this utterance results in something perfectly understandable.

8. For a discussion of modeling within the philosophy of science, see Bailer-Jones (2009).

9. See McShane et al. (2008) and McShane and Nirenburg (2021) for details.

10. This is an arbitrary large number that signifies "never."

11. These are redrawn versions of the screenshots for higher quality. More screenshots are available at https://faculty.rpi.edu/marjorie-mcshane.

Chapter 9

1. From the article "For AI, data are harder to come by than you think," June 13, 2020, *The Economist*. https://www.economist.com/technology-quarterly/2020/06/11/for-ai-data-are-harder-to-come-by-than-you-think

2. For discussion, see McShane and Nirenburg (2021), section 2.8.1.

3. McShane and Nirenburg (2021, ch. 8) provide an extended example of cognitive modeling in the medical domain, as well as the capabilities such models enable in cognitive agent systems.

4. Specifically, MEAL in the ontology has the children BREAKFAST, LUNCH, DINNER, SNACK, to which the words *breakfast*, *lunch*, *dinner*, and *tea* map, respectively.

5. For a lexically oriented classification and literature review, see Bhagat and Hovy (2013).

6. Arguably, there are few true synonyms in language—thus the term *near-synonym* (Hirst, 1995).

7. There will still remain genuine cases of residual ambiguity that require contextual reasoning to resolve, such as direct and metaphorical senses of verbs: for example, *to pummel* can mean *to beat up* or *to criticize severely*. But careful acquisition can greatly improve automatic disambiguation.

8. See McShane and Nirenburg (2021), section 2.7.

9. For a recent survey, see Althunayyan and Azmi (2021).

10. See McShane and Nirenburg (2021), section 5.4.

11. See the multiple discussions of unexpected input in McShane and Nirenburg (2021).

Chapter 10

1. See Herbert Roitblat's (2023) "Does artificial intelligence threaten human extinction?," accessed on June 22, 2023 at https://bdtechtalks.com/2023/06/15/artificial-intelligence-human-extinction/ and Kathleen Miles' (2014) "Artificial intelligence may doom the human race within a century, Oxford professor says," accessed on June 22, 2023 at https://www.huffpost.com/entry/artificial-intelligence-oxford_n_5689858.

2. "AI is a lot of work," *New York Magazine*, June 20, 2023. Accessed at https://nymag.com/intelligencer/article/ai-artificial-intelligence-humans-technology-business-factory.html on June 23, 2023.

3. "Talking about AI in human terms is natural—but wrong," *The Economist*, June 22, 2023, https://www.economist.com/culture/2023/06/22/talking-about-ai-in-human-terms-is-natural-but-wrong.

4. "Talking about AI in human terms is natural—but wrong," *The Economist*, June 22, 2023, https://www.economist.com/culture/2023/06/22/talking-about-ai-in-human-terms-is-natural-but-wrong.

References

Albert, T. (2004). Why are medical journals so badly written? *Medical Education*, *38*(1), 6–8.

Althunayyan, S., & Azmi, A. (2021). Automated text simplification: A survey. *ACM Computing Surveys*, *54*(2), 1–36. Article no. 43. https://doi.org/10.1145/3442695

Argall, B., Chernova, S., Veloso, M. M., & Browning, B. (2009). A survey of robot learning from demonstration. *Robotics & Autonomous Systems*, *57*, 469–483.

Babic, B., Gerke, S., Evgeniou, T., & Cohen, I. G. (2021). Beware explanations from AI in health care. *Science*, *373*(6552), 284–286.

Bailer-Jones, D. M. (2009). *Scientific models in philosophy of science*. University of Pittsburgh Press.

Barker, K., Agashe, B., Chaw, S., Fan, J., Friedland, N., Glass, N. M., Hobbs, J., Hovy, E., Israel, D., Kim, D. S., Multar-Mehta, R., Patwardhani, S., Porter, B., Tecuci, D., & Yeh, P. (2007). Learning by reading: A prototype system, performance baseline and lessons learned. *Proceedings of the Twenty-Second Conference on Artificial Intelligence*, 280–286. Association for the Advancement of Artificial Intelligence.

Barredo Arrieta, A., Díaz-Rodríguez, N., Del Ser, J., Bennetot, A., Tabik, S., Barbado, A., Garcia, S., Gil-Lopez, S., Molina, D., Benjamins, R., Chatila, R., & Herrera, F. (2020). Explainable artificial intelligence (XAI): Concepts, taxonomies, opportunities and challenges toward responsible AI. *Information Fusion*, *548*, 82–115.

Beatty, J. (1982). Task-evoked pupillary responses, processing load, and the structure of processing resources. *Psychological Bulletin*, *91*(2), 276–292. https://doi.org/10.1037/0033-2909.91.2.276

Bengio, Y. (2019). From System 1 deep learning to System 2 deep learning (NeurIPS 2019). A lecture available at https://www.youtube.com/watch?v=T3sxeTgT4qc&t=3s.

Bergen, B., & Chang, N. (2013). Embodied construction grammar. In T. Hoffmann & G. Trousdale, (Eds.). *The Oxford handbook of construction grammar* (pp. 168–190). Oxford University Press.

Bhagat, R., & Hovy, E. (2013). What is a paraphrase? *Computational Linguistics*, *39*(3), 463–472. Association for Computational Linguistics.

Boguslavsky, I. M. (2021). Semantičeskij analiz s oporoj na umozaključenija v funkcional'noj modeli jazyka. *Voprosy jazykoznanija*, 1, 29–56.

Brachman, R. J., & Levesque, H. J. (2022). *Machines like us: Toward AI with common sense*. MIT Press.

Bringsjord, S., Giancola, M., & Govindarajulu, N. S. (2023). Logic-based modeling of cognition. In R. Sun (Ed.), *The Cambridge handbook of computational cognitive sciences* (pp. 173–209). Cambridge University Press.

Brown, T., Mann, B., Ryder, N., Subbiah, M., Kaplan, J. D., Dhariwal, P., Neelakantan, A., Shyam, P., Sastry, G., Askell, A., Agarwal, S., Herbert-Voss, A., Krueger, G., Henighan, T., Child, R., Ramesh, A., Ziegler, D. M., Wu, J., Winter, C., . . . Amodei, D. (2020). Language models are few-shot learners. *Advances in Neural Information Processing Systems*, *33*, 1877–1901.

Bryant, J. E. (2008). *Best-fit constructional analysis*. Technical Report No. UCB/EECS-2008–100. Electrical Engineering and Computer Sciences, University of California at Berkeley. https://www2.eecs.berkeley.edu/Pubs/TechRpts/2008/EECS-2008-100.html

Bubeck, S., Chandrasekaran, V., Eldan, R., Gehrke, J., Horvitz, E., Kamar, E., Lee, P., Tat Lee, Y., Li, Y., Lundberg, S., Nori, H., Palangi, H., Tulio Ribeiro, M., & Zhang, Y. (2023). Sparks of artificial general intelligence: Early experiments with GPT-4. *arXiv*. https://arxiv.org/abs/2303.12712

Burrows, S., Potthast, M., & Stein, B. (2013). Paraphrase acquisition via crowdsourcing and machine learning. *Transactions on Intelligent Systems and Technology*, *4*(3), 1–21.

Bybee, J. L. (2013). Usage-based theory and exemplar representations of constructions. In T. Hoffmann & G. Trousdale (Eds.), *The Oxford handbook of construction grammar* (pp. 49–69). Oxford University Press.

Carlson, L., Marcu, D., & Okurowski, M. E. (2003). Building a discourse-tagged corpus in the framework of Rhetorical Structure Theory. In J. van Kuppevelt & R. W. Smith (Eds.), *Current and new directions in discourse and dialogue* (pp. 85–112). Kluwer.

Cartwright, N. (1983). *How the laws of physics lie*. Oxford University Press.

Chan, S., & Siegel, E. L. (2019). Will machine learning end the viability of radiology as a thriving medical specialty? *British Journal of Radiology*, *92*(1094). https://doi.org/10.1259/bjr.20180416

Charniak, E. (1972). *Toward a model of children's story comprehension* [Unpublished doctoral dissertation]. Massachusetts Institute of Technology.

Chomsky, N. (1965). *Aspects of the theory of syntax*. MIT Press.

Chomsky, N. (1995). *The minimalist program*. MIT Press.

Chowdhery, A., Narang, S., Devlin, J., Bosma, M., Mishra, G., Roberts, A., Barham, P., Chung, H. W., Sutton, C., Gehrmann, S., Schuh, P., Shi, K., Tsvyashchenko, S., Maynez, J., Rao, A., Barnes, P., Tay, Y., Shazeer, N., Prabhakaran, V., . . . Fiedel, N. (2022). Palm: Scaling language modeling with path-ways. *arXiv:2204.02311*.

Christiansen, M. H., & Chater, N. (2016). The now-or-never bottleneck: A fundamental constraint on language. *Behavioral and Brain Sciences, 39*, 1–72.

Clark, S. (2015). Vector space models of lexical meaning. In S. Lappin & C. Fox (Eds.), *The handbook of contemporary semantic theory* (2nd ed., pp. 493–522). Wiley.

Culicover, P. W., & Jackendoff, R. (2005). *Simpler syntax*. Oxford University Press.

Dale, R. (1989). Cooking up referring expressions. *Proceedings of the 27th Annual Meeting of the Association for Computational Linguistics*. Association for Computational Linguistics.

Dale, R., & Reiter, E. (1995). Computational interpretations of the Gricean maxims in the generation of referring expressions. *Cognitive Science, 19*, 233–263.

Das, D., & Taboada, M. (2018). RST signalling corpus: A corpus of signals of coherence relations. *Language Resources and Evaluation, 52*(1), 149–184. https://doi.org/10.1007/s10579-017-9383-x

Davies, M. (2008–). The Corpus of Contemporary American English (COCA): One billion words, 1990–2019. https://www.english-corpora.org/coca/

Deemter, K. v., Gatt, A., Sluis, I. v. d., & Power, R. (2012). Generation of referring expressions: Assessing the Incremental Algorithm. *Cognitive Science, 36*(5), 799–836.

Dickson, B. (2019, December 23). System 2 deep learning: The next step toward artificial general intelligence. *TechTalks*. https://bdtechtalks.com/2019/12/23/yoshua-bengio-neurips-2019-deep-learning/

Diessel, H. (2016). Frequency and lexical specificity in grammar: A critical review. In H. Behrens & S. Pfänder (Eds.), *Experience counts: Frequency effects in language*, (vol. 54, pp. 208–238; in the series Linguae & Litterae). De Gruyter, Inc.

Dolan, B., Brockett, C., & Quirk, C. (2005). Microsoft research paraphrase corpus. The "readme" in the corpus is available at https://www.microsoft.com/en-us/download/details.aspx?id=52398.

Dwivedi, R., Dave, D., Naik, H., Singhal, S., Omer, R., Patel, P., Qian, B., Wen, Z., Shah, T., Morgan, G., & Ranjan, R. (2022). Explainable AI (XAI): Core ideas, techniques and solutions. *ACM Computing Surveys, 55*(9), 1–33. https://doi.org/10.1145/3561048

Ehsan, U., Wintersberger, P., Liao, Q. V., Watkins, E. A., Manger, C., Daumé, Hal, III, Riener, A., & Riedl, M. O. (2022). Human-centered explainable AI (HCXAI): Beyond opening the black-box of AI. *CHI EA '22: Extended abstracts of the 2022 CHI Conference on Human Factors in Computing Systems*, pp. 1–7. Association for Computing Machinery. https://doi.org/10.1145/3491101.3503727

Elder, R., & Krishna, A. (2014). Grasping the grounded nature of mental simulation. *The Inquisitive Mind, 4*(20). https://www.in-mind.org/article/grasping-the-grounded -nature-of-mental-simulation

English, J. (2010). *Learning by reading: Automatic knowledge extraction through semantic analysis* [Unpublished doctoral dissertation]. University of Maryland Baltimore County.

English, J., & Nirenburg, S. (2007). Ontology learning from text using automatic ontological-semantic annotation and the web as a corpus. *Proceedings of the AAAI Spring Symposium on Machine Reading*, pp. 43–48. AAAI Press.

English, J., & Nirenburg, S. (2019). XMRs: Uniform semantic representations for intermodular communication in cognitive robots. *Proceedings from the Seventh Annual Conference on Advances in Cognitive Systems Workshop on Cognitive Robotics*. Cognitive Systems Foundation.

Eppe, M., Trott, S., & Feldman, J. (2016). Exploiting deep semantics and compositionality of natural language for human-robot-interaction. *2016 IEEE/RSJ International Conference on Intelligent Robots and Systems*, pp. 731–738. IEEE. https://doi.org /10.1109/IROS.2016.7759133

Eppe, M., Trott, S., Raghuram, V., Feldman, J., & Janin, A. (2016). Application-independent and integration-friendly natural language understanding. In C. Benzmüller, G. Sutcliffe, & R. Rojas (Eds.), *GCAI 2016. 2nd Global Conference on Artificial Intelligence* (vol. 41, pp. 340–352). EasyChair. https://doi.org/10.29007/npsn

Erol, K., Hendler, J., & Nau, D. S. (1994). HTN planning: Complexity and expressivity. *Proceedings of the Twelfth National Conference on Artificial Intelligence*, pp. 1123–1128. AAAI Press.

Feldman, J. A. (2020). Advances in embodied construction grammar. *Constructions and Frames, 12*(1), 149–169. https://doi.org/10.1075/cf.00038.fel

Fillmore, C. J. (1985). Frames and the semantics of understanding. *Quaderni di Semantica, 6*(2), 222–254.

Fillmore, C. J., & Baker, C. F. (2009). A frames approach to semantic analysis. In B. Heine & H. Narrog (Eds.), *The Oxford handbook of linguistic analysis* (pp. 313–340). Oxford University Press.

Fillmore, C. J., Lee-Goldman R., & Rhodes R. (2012). The FrameNet constructicon. In H. C. Boas & I. A. Sag (Eds.), *Sign-Based Construction Grammar* (pp. 309–372). Center for the Study of Language and Information.

Firth, J. R. (1957). A synopsis of linguistic theory, 1930–1955. In J. R. Firth (Ed.), *Studies in linguistic analysis* (pp. 1–32). Blackwell. (Reprinted in *Selected papers of J. R. Firth 1952–1959*, by F. R. Palmer, Ed., 1968, Longman).

Forbus, K. D. (1984). Qualitative process theory. *Artificial Intelligence, 24*(1–3), 85–168.

Forbus, K. D. (2019). *Qualitative representations.* MIT Press.

Forbus, K., Lockwood, K., Tomai, E., Dehghani, M., & Czyz, J. (2007). Machine reading as a cognitive science research instrument. *Proceedings of the AAAI Spring Symposium on Machine Reading.* Association for the Advancement of Artificial Intelligence.

Fox, J. J. (1977). Roman Jakobson and the comparative study of parallelism. In C. H. van Schooneveld & D. Armstrong (Eds.), *Roman Jakobson: Echoes of his scholarship* (pp. 59–90). Peter de Ridder Press.

Frigg, R., & Hartman, S. (2020). Models in science. In E. N. Zalta (Ed.), *The Stanford encyclopedia of philosophy* (Spring 2020 Edition). https://plato.stanford.edu/archives/spr2020/entries/models-science

Friedman, T. (2023, March 21). Our new Promethian moment. *The New York Times.*

Gatt, A., & Krahmer, E. (2018). Survey of the state of the art in natural language generation: Core tasks, applications and evaluation. *Journal of Artificial Intelligence Research, 61*(c), 1–64.

Gatt, A., & Reiter, E. (2009). SimpleNLG: A realization engine for practical applications. *Proceedings of the 12th European Workshop on Natural Language Generation,* pp. 90–93. Association for Computational Linguistics.

Gigerenzer, G. (2008). Why heuristics work. *Perspectives on Cognitive Science, 3*(1), 20–29.

Godfrey, J., Holliman, E., & McDaniel, J. (1992). Switchboard: Telephone speech corpus for research and development. *Proceedings of the 1992 IEEE International Conference on Acoustics, Speech and Signal Processing* (vol. 1, pp. 517–520). IEEE.

Goldberg, A. E. (2006). *Constructions at work: The nature of generalizations in language.* Oxford University Press.

Goldberg, A. E. (2013). Constructionist approaches. In T. Hoffmann & G. Trousdale (Eds.), *The Oxford handbook of construction grammar* (pp. 15–31). Oxford University Press.

Goldberg, A. E. (2019). *Explain me this: Creativity, competition, and the partial productivity of constructions.* Princeton University Press.

Goldwater, S., Jurafsky, D., & Manning, C. D. (2010). Which words are hard to recognize? Prosodic, lexical, and disfluency factors that increase speech recognition error rates. *Speech Communication, 52,* 181–200.

Goodall, G. (1987). *Parallel structures in syntax: Coordination, causatives, and restructuring*. Cambridge University Press.

Goyal, K., Jauhar, S. K., Li, H., Sachan, M., Srivastava, S., & Hovy, E. (2013). A structured distributional semantic model: Integrating structure with semantics. *Proceedings of the Workshop on Continuous Vector Space Models and Their Compositionality*, pp. 20–29. The Association for Computational Linguistics.

Graff, D., & Cieri, C. (2003). *English Gigaword* (LDC2003T05). Linguistic Data Consortium. https://catalog.ldc.upenn.edu/LDC2003T05

Gray, T. (2019). *How things work: The inner life of everyday machines*. Hachette UK.

Gunning, D. (2017). Explainable artificial intelligence (XAI). DARPA/I20. Program Update November 2017. https://nsarchive.gwu.edu/sites/default/files/documents/5794867/National-Security-Archive-David-Gunning-DARPA.pdf

Hakes, T. (2021, March 23). How much work can you really get done? *7Pace blog*. https://www.7pace.com/blog/how-much-work-can-you-really-get-done

Hardcastle, V. G., & Hardcastle, K. (2015). Marr's levels revisited: Understanding how brains break. *Topics in Cognitive Science, 7*(2), 259–273.

Hayakawa, S. I. (Ed.). (1994). *Choose the right word* (2nd ed., revised by Eugene Ehrlich). Harper Collins Publishers.

Hayes, P. J. (1979). The naive physics manifesto. In D. Michie (Ed.), *Expert systems in the micro-electronic age* (pp. 242–270). Edinburgh University Press.

Healey, P. G., Howes, C., Hough, J., & Purver, M. (2016). Better late than now-or-never: The case of interactive repair phenomena. *The Behavioral and Brain Sciences, 39*, e76. https://doi.org/10.1017/S0140525X15000813

Heuer, R. J., Jr. (1999). *Psychology of intelligence analysis*. Center for the Study of Intelligence, Central Intelligence Agency. https://www.cia.gov/static/Pyschology-of-Intelligence-Analysis.pdf

Hirschman, L., & Chinchor, N. (1997). MUC-7 coreference task definition (Version 3.0). *Proceedings of the Seventh Message Understanding Conference*. The Association for Computational Linguistics.

Hirst, G. (1995). Near-synonymy and the structure of lexical knowledge. *Proceedings of the AAAI Symposium on Representation and Acquisition of Lexical Knowledge: Polysemy, Ambiguity, and Generativity*, pp. 51–56. AAAI Press.

Hobbs, J., & Kehler, A. (1998). A theory of parallelism and the case of VP ellipsis. *Proceedings of the 36th Annual Meeting of the Association for Computational Linguistics*, pp. 394–401. Association for Computational Linguistics.

Hoffmann, T., & Trousdale, G. (Eds.). (2013). *The Oxford handbook of construction grammar*. Oxford University Press.

Honnibal, M., & Montani, I. (2017). spaCy 2: Natural language understanding with Bloom embeddings, convolutional neural networks and incremental parsing. https://spacy.io/

Ide, N., & Véronis, J. (1993). Extracting knowledge bases from machine-readable dictionaries: Have we wasted our time? *Proceedings of the Workshop from the 1st Conference and Workshop on Building and Sharing of Very Large-Scale Knowledge Bases*, pp. 257–266. AI Communications.

Inkpen, D. Z., & Hirst, G. (2003). Automatic sense disambiguation of the near-synonyms in a dictionary entry. *Proceedings of the Fourth International Conference on Computational Linguistics and Intelligent Text Processing* (pp. 258–267). Lecture Notes in Computer Science. Springer-Verlag.

Jacobsson, H., Kruijff, G-J., Hawes, N., & Wyatt, J. (2007). Crossmodal content binding in information-processing architectures. *Proceedings of the Third Human-Robot Interaction Conference*, pp. 81–88. IEEE. https://doi.org/10.1145/1349822.1349834

Jakobson, R., & Vine, B. (1985). Poetry of grammar and grammar of poetry. In K. Pomorska & S. Rudy (Eds.), *Verbal art, verbal sign, verbal time* (pp. 37–46). University of Minnesota Press.

Jeong, M., & Lee, G. G. (2006). Jointly predicting dialog act and named entity for spoken language understanding. *Proceedings of the IEEE Spoken Language Technology Workshop*, pp. 66–69. IEEE.

Jurafsky, D. (2006). Pragmatics and computational linguistics. In L. R. Horn & G. Ward (Eds.), *The handbook of pragmatics*. Blackwell Publishing. https://doi.org/10.1002/9780470756959.ch26

Jurafsky, D., & Martin, J. H. (2023). *Speech and language processing* (3rd ed.). [Unpublished manuscript]. Draft of January 7, 2023. https://web.stanford.edu/~jurafsky/slp3/

Kaddour, J., Harris, J., Mozes, M., Bradley, H., Raileanu, R., & McHardy, R. (2023). Challenges and applications of large language models. *arXiv:2307.10169 [cs.CL]*.

Kahneman, D. (2011). *Thinking: Fast and slow*. Farrar, Strauss and Giroux.

Kahneman, D., & Beatty, J. (1966). Pupil diameter and load on memory. *Science, 154*(3756), 1583–1585. http://www.jstor.org/stable/1720478

Kaur, D., Uslu, S., Rittichier, K. J., & Durresi, A. (2022). Trustworthy artificial intelligence: A review. *ACM Computing Surveys, 55*, 1–38.

Kay, P. (2013). The limits of (construction) grammar. In T. Hoffmann & G. Trousdale (Eds.), *The Oxford handbook of construction grammar* (pp. 32–48). Oxford University Press.

Keizer, E. (2014). Context and cognition in Functional Discourse Grammar: What, where and why? *Pragmatics, 24*(2), 399–423.

Kilgarriff, A., Baisa, V., Bušta, J., Jakubíček, M., Kovář, V., Michelfeit, J., Rychlý, P., & Suchomel, V. (2014). The Sketch Engine: Ten years on. *Lexicography, 1*, 7–36.

Kitaev, N., Cao, S., & Klein, D. (2019). Multilingual constituency parsing with self-attention and pre-training. *Proceedings of the 57th Annual Meeting of the Association for Computational Linguistics*, pp. 3499–3505. Association for Computational Linguistics. https://doi.org/10.18653/v1/P19-1340

Korte, R. F. (2003). Biases in decision making and implications for human resource development. *Advances in Developing Human Resources, 5*(4), 440–457.

Kotseruba, I., & Tsotsos, J. K. (2018). A review of 40 years in cognitive architecture research: Core cognitive abilities and practical applications. *arXiv:1610.08602*.

Král, P., & Cerisara, C. (2010). Dialogue act recognition approaches. *Computing & Informatics, 29*, 227–250.

Krishnaswamy, N., & Pustejovsky, J. (2016). VoxSim: A visual platform for modeling motion language. *Proceedings of the 26th International Conference on Computational Linguistics* (COLING 2016), pp. 54–58. The COLING 2016 Organizing Committee.

Laird, J. E. (2012). *The SOAR cognitive architecture*. MIT Press.

Langley, P., Laird, J. E., & Rogers, S. (2009). Cognitive architectures: Research issues and challenges. *Cognitive Systems Research, 10*, 141–160.

Lee, K., He, L., Lewis, M., & Zettlemoyer, L. (2017). End-to-end neural coreference resolution. *Proceedings of the 2017 Conference on Empirical Methods in Natural Language Processing*, pp. 188–197. Association for Computational Linguistics.

Lenat, D. B., Gula, R., Pittman, K., Pratt, D., & Shepherd, M. (1990). CYC: Toward programs with common sense. *Communications of the ACM, 33*(8), 30–49. https://doi.org/10.1145/79173.79176

Lenat, D., Miller, G., & Yokoi, T. (1995). CYC, WordNet, and EDR: Critiques and responses. *Communications of the ACM, 38*(11), 45–48. https://doi.org/10.1145/219717.219757

Lenat, D. (2019, February 18). Not good as gold: Today's AIs are dangerously lacking in AU (artificial understanding). *Forbes*.

Levesque, H., Davis, E., & Morgenstern, L. (2012). The Winograd Schema Challenge. *Proceedings of the Thirteenth International Conference on Principles of Knowledge Representation and Reasoning*, pp. 552–561. AAAI Press.

Levin, B. (1993). *English verb classes and alternations: A preliminary investigation*. University of Chicago Press.

Levinson, S. C. (2016). "Process and perish" or multiple buffers with push-down stacks? *The Behavioral and Brain Sciences, 39*, e81. https://doi.org/10.1017/S0140525X15000862

Lewis, M. L., & Frank, M. C. (2016). Linguistic structure emerges through the interaction of memory constraints and communicative pressures. *The Behavioral and Brain Sciences, 39*, e82. https://doi.org/10.1017/S0140525X15000874

Liao, V. Q., Pribić, M., Han, J., Miller, S., & Sow, D. (2021). Question-driven design process for explainable AI user experiences. *arXiv: 2104.03483.*

Liao, V. Q., & Varshney, K. R. (2022). Human-centered explainable AI (XAI): From algorithms to user experiences. *arXiv:2110.10790.*

Lieto, A., Lebiere, C., & Oltramari, A. (2018). The knowledge level in cognitive architectures: Current limitations and possible developments. *Cognitive Systems Research, 48*, 39–55.

Lohr, S. (2021, July 16). What ever happened to IBM's Watson? *The New York Times.* Updated July 17, 2021.

Lu, J., & Ng, V. (2018). Event coreference resolution: A survey of two decades of research. *Proceedings of the 27th International Joint Conference on Artificial Intelligence,* pp. 5479–5486. AAAI Press.

Lyngfelt, B., Borin, L., Ohara, K., & Torrent, T. T. (2018). *Constructicography: Constructicon development across languages.* John Benjamins Publishing Company.

MacDonald, M. C. (2016). Memory limitations and chunking are variable and cannot explain language structure. *The Behavioral and Brain Sciences, 39*, e84. https://doi.org/10.1017/S0140525X15000898

Mahesh, K., Nirenburg, S., Farwell, D, & Cowie, J. (1996). An assessment of CYC for natural language processing. Technical Report, Computing Research Lab, New Mexico State University. https://www.researchgate.net/publication/277296502_An_Assessment_of_CYC_for_Natural_Language_Processing

Mahowald, K., Ivanova, A. A., Blank, I. A, Kanwisher, N., Tenenbaum, J. B., & Fedorenko, E. (2023). Dissociating language and thought in large language models: A cognitive perspective. *arXiv:2301.06627.*

Malle, B. (2004). *How the mind explains behavior: Folk explanations, meaning and social interaction.* MIT Press.

Marchal, M., Scholman, M. C. J., & Demberg, V. (2020). The effect of domain knowledge on discourse relation inferences: Relation marking and interpretation strategies. *Dialogue & Discourse, 13*(2), 49–78. https://doi.org/10.5210/dad.2022.202

Marcus, G. (2022, March 10). Deep learning is hitting a wall. *Nautilus.* https://nautil.us/deep-learning-is-hitting-a-wall-14467

Marr, D. (1982). *Vision: A computational approach.* W. H. Freeman.

Marín-Morales, J., Llinares, C., Guixeres, J., & Alcañiz, M. (2020). Emotion recognition in immersive virtual reality: From statistics to affective computing. *Sensors, 20*(5163), 1–26. https://doi.org/doi:10.3390/s20185163

Mazumder, S., Ma, N., and Liu, B. (2018). Towards a continuous knowledge learning engine for chatbots. *arXiv:1802.06024*.

Matzkin, A. (2021, September 29). AI in Healthcare: Insights from two decades of FDA approvals. *Health Advances blog*. https://healthadvances.com/insights/blog/ai-in -healthcare-insights-from-two-decades-of-fda-approvals

McDonald, David D. (1983). Description directed control: Its implications for natural language generation. In N. Brady (Ed.), *Computational linguistics* (pp. 111–129). Pergamon Press.

McKinsey Analytics. (2021, December). The state of AI in 2021. https://www .mckinsey.com/~/media/McKinsey/Business%20Functions/McKinsey%20Analytics /Our%20Insights/Global%20survey%20The%20state%20of%20AI%20in%202021 /Global-survey-The-state-of-AI-in-2021.pdf

McShane, M. (2000). Hierarchies of parallelism in elliptical Polish structures. *Journal of Slavic Linguistics*, *8*, 83–117.

McShane, M. (2005). *A theory of ellipsis*. Oxford University Press.

McShane, M. (2014). Parameterizing mental model ascription across intelligent agents. *Interaction Studies*, *15*(3), 404–425.

McShane, M. (2018). Typical event sequences as licensors of direct object ellipsis in Russian. *Lingvisticæ Investigationes*, *41*(2), 179–212.

McShane, M., & Babkin, P. (2016a). Detection and resolution of verb phrase ellipsis. *Linguistic Issues in Language Technology*, *13*(1), 1–34.

McShane, M., & Babkin, P. (2016b). Resolving difficult referring expressions. *Advances in Cognitive Systems*, *4*, 247–263.

McShane, M., English, J., & Nirenburg, S. (2021). Knowledge engineering in the long game of artificial intelligence: The case of speech acts. *Proceedings of the Ninth Annual Conference on Advances in Cognitive Systems*. Cognitive Systems Foundation.

McShane, M., Jarrell, B., Fantry, G., Nirenburg, S., Beale, S., & Johnson, B. (2008). Revealing the conceptual substrate of biomedical cognitive models to the wider community. In J. D. Westwood, R. S. Haluck, H. M. Hoffman, G. T. Mogel, R. Phillips, R. A. Robb, & K. G. Vosburgh (Eds.), *Medicine meets virtual reality 16: Parallel, combinatorial, convergent: NextMed by design* (pp. 281–286). IOS Press.

McShane, M., & Leon, I. (2021). Language generation for broad-coverage, explainable cognitive systems. *Proceedings of the Ninth Annual Conference on Advances in Cognitive Systems*. Cognitive Systems Foundation.

McShane, M., & Nirenburg, S. (2012). A knowledge representation language for natural language processing, simulation and reasoning. *International Journal of Semantic Computing*, *6*(1), 3–23.

McShane, M., & Nirenburg. S. (2021). *Linguistics for the age of AI*. MIT Press. Available, open access, at https://direct.mit.edu/books/book/5042/Linguistics-for-the-Age-of-AI.

McShane, M., & Nirenburg, S. (2023). Natural language understanding and generation. In R. Sun (Ed.), *The Cambridge handbook of computational cognitive sciences* (chapter 28, pp. 921–946). Cambridge University Press.

McShane, M., Nirenburg, S. & Beale, S. (2005). An NLP lexicon as a largely language independent resource. *Machine Translation, 19*(2), 139–173.

McShane, M., Nirenburg, S., Cowie, J., & Zacharski, R. (2002). Embedding knowledge elicitation and MT systems within a single architecture. *Machine Translation, 17*(4), 271–305.

McShane, M., Nirenburg, S., Jarrell, B., & Fantry, G. (2015). Learning components of computational models from texts. In M. A. Finlayson, B. Miller, A. Lieto, & R. Ronfard (Eds.), *Proceedings of the 6th Workshop on Computational Models of Narrative* (pp. 108–123). Published in the Open Access Series in Informatics [OASIcs], Schloss Dagstuhl—Leibniz-Zentrum für Informatik, Dagstuhl Publishing, Germany.

Miller, G.A. (1995). Wordnet: A lexical database for English. *Communications of the ACM, 38*, 39–41.

Minsky, M. (1975). A framework for representing knowledge. In P. Winston (Ed.), *The psychology of computer vision*. McGraw-Hill.

Minsky, M. (2006). *The emotion machine*. Simon and Schuster.

Mitchell, M. (2023). On detecting whether text was generated by a human or an AI language model. https://aiguide.substack.com/p/on-detecting-whether-text-was-generated

Mitchell, T., Cohen, W., Hruschka, E., Talukdar, P., Yang, B., Betteridge, J., Carlson, A., Dalvi, B., Gardner, M., Kisiel, B., Krishnamurthy, J., Lao, N., Mazaitis, K., Mohamed, T., Nakashole, N., Platanios, E., Ritter, A., Samadi, M., Settles, B., . . . Welling, J. (2018). Never-ending learning. *Communications of the ACM, 61*(5), 103–115. https://doi.org/10.1145/3191513

Mueller, S. T., Hoffman, R. R., Clancey, W., Emrey, A., & Klein, G. (2019). Explanation in human-AI systems: A literature meta-review. Synopsis of key ideas and publications, and bibliography for explainable AI. Technical Report, DARPA Explainable AI Program. *arXiv:1902.01876*.

Nunnally, J. C., Knott, P. D., Duchnowski, A., & Parker, R. (1967). Pupillary response as a general measure of activation. *Perception & Psychophysics, 2*, 149–155.

Nauta, M., Trienes, J., Pathak, S., Nguyen, E., Peters, M., Schmitt, Y., Shlötterer, J., van Keulen, M., & Seifert, C. (2023). From anecdotal evidence to quantitative evaluation methods: A systematic review on evaluating explainable AI. *arXiv:2201.08164v3*.

Newell, A. (1982). The knowledge level. *Artificial Intelligence, 18*, 87–127.

Newmeyer, F. J., & Preston, L. B. (2014). *Measuring grammatical complexity*. Oxford University Press.

Nirenburg, S. (2017). Cognitive systems: Toward human-level functionality. Special Issue on Cognitive Systems. *Artificial Intelligence Magazine, 38*(4), 5–12. https://doi .org/10.1609/aimag.v38i4.2760

Nirenburg, S., English, J., & McShane, M. (2021). Artificial intelligent agents go to school. *34th International Workshop on Qualitative Reasoning at IJCAI-21*. https://www .qrg.northwestern.edu/qr2021/papers/QR2021_NirenburgEnglishMcShane.pdf

Nirenburg, S., Krishnaswamy, N., & McShane, M. (2023). Hybrid ML/KB systems learning through NL dialog with DL models. In A. Martin, K. Hinkelmann, H.-G. Fill, A. Gerber, D. Lenat, R. Stolle, & F. van Harmelen (Eds.), *Proceedings of the AAAI 2023 Spring Symposium on Challenges Requiring the Combination of Machine Learning and Knowledge Engineering*. AAAI Press.

Nirenburg, S., McShane, M., & Beale, S. (2009). A unified ontological-semantic substrate for physiological simulation and cognitive modeling. *Proceedings of the International Conference on Biomedical Ontology* (ICBO-2009), pp. 139–142.

Nirenburg, S., McShane, M., Beale, S., Wood, P., Scassellati, B., Mangin, O., & Roncone, A. (2018). Toward human-like robot learning. *Proceedings of the Twenty Third International Conference on Natural Language and Information Systems*, pp. 73–82.

Nirenburg, S., McShane, M., & English, J. (2023). Content-centric computational cognitive modeling. *Advances in Cognitive Systems*.

Nirenburg, S., Oates, T., & English, J. (2007). Learning by reading by learning to read. *Proceedings of the International Conference on Semantic Computing*, pp. 651–658.

Nirenburg, S., & Raskin, V. (2004). *Ontological semantics*. MIT Press.

Nirenburg, S., Somers, H., & Wilks, Y. (2003). *Readings in machine translation*. MIT Press.

Nirenburg, S., & Wood, P. (2017). Toward human-style learning in robots. *AAAI Fall Symposium on Natural Communication for Human-Robot Collaboration*.

O'Hara, T., & Wiebe, J. (2009). Exploiting semantic role resources for preposition disambiguation. *Computational Linguistics, 35* (2), 151–184.

Palmer, M., Gildea, D., & Kingsbury, P. (2005). The proposition bank: An annotated corpus of semantic roles. *Computational Linguistics, 31*(1), 71–105.

Pantel, P., Ravichandran, D., & Hovy, E. (2004). Towards terascale knowledge acquisition. *Proceedings of the 20th International Conference on Computational Linguistics*, pp. 771–777. COLING. https://doi.org/10.3115/1220355.1220466

Parisi, G. I., Kemker, R., Part, J. L., Kanan, C., & Wermter, S. (2019). Continual life-long learning with neural networks: A review. *Neural Networks, 113*, 54–71.

Pinker, S. (2014, September 26). Why academics stink at writing. *The Chronicle of Higher Education*. https://www.chronicle.com/article/why-academics-stink-at-writing/

Poesio, M., Stuckardt, R., & Versley, Y. (Eds.). (2016). *Anaphora resolution: Algorithms, resources, and applications*. Springer-Verlag.

Pustejovsky, J., & Krishnaswamy, N. (2016). Voxml: A visualization modeling language. *Proceedings of the Tenth International Conference on Language Resources and Evaluation* (LREC'16), pp. 4606–4613. European Language Resources Association.

Reiter, E. (1990). The computational complexity of avoiding conversational implicatures. *Proceedings of the 28th Annual Meeting of the Association for Computational Linguistics* (ACL '90), pp. 97–104. Association for Computational Linguistics.

Roncone, A., Mangin, O., & Scassellati, B. (2017). Transparent role assignment and task allocation in human robot collaboration. *2017 IEEE International Conference on Robotics and Automation*, pp. 1014–1021. https://doi.org/10.1109/ICRA.2017.7989122

Rosenberg, A., & Hasegawa-Johnson, M. (2020). Automatic prosody labelling and assessment. In C. Gussenhoven & A. Chen (Eds.), *The Oxford handbook of language prosody* (pp. 646–656). Oxford University Press.

Santhanam, S., & Shaikh, S. (2019). A survey of natural language generation techniques with a focus on dialogue systems—Past, present and future directions. *arXiv:1906.00500*.

Scao, T. L., Fan, A., Akiki, C., Pavlick, E., Ilić, S., Hesslow, D., Castagné, R., Luccioni, A.S., Yvon, F., Gallé, M., & Tow, J. (2022). Bloom: A 176b-parameter open-access multilingual language model. *arXiv:2211.05100*.

Schank, R. (1982). *Dynamic memory: A theory of learning in computers and people*. Cambridge University Press.

Schank, R., & Abelson, R. P. (1977). *Scripts, plans, goals and understanding: An inquiry into human knowledge structures*. Erlbaum.

Scheutz, M., Harris, J., & Schermerhorn, P. (2013). Systematic integration of cognitive and robotic architectures. *Advances in Cognitive Systems, 2*, 277–296. Cognitive Systems Foundation.

Scheutz, M., Krause, E., Oosterveld, B., Frasca, T., & Platt, R. (2017). Spoken instruction-based one-shot object and action learning in a cognitive robotic architecture. In S. Das, E. Durfee, K. Larson, & M. Winikoff (Eds.), *Proceedings of the 16th International Conference on Autonomous Agents and Multiagent Systems*. International Foundation for Autonomous Agents and Multiagent Systems.

Simon, H. (1957). *Models of man, social and rational: Mathematical essays on rational human behavior in a social setting.* Wiley.

Sloman, A., & Scheutz, M. (2002). A framework for comparing agent architectures. In J. A. Bullinaria (Ed.), *Proceedings of the 2002 UK Workshop on Computational Intelligence.* The University of Birmingham.

Smiley, L. (2022, March 8). 'I'm the operator': The aftermath of a self-driving tragedy. *Wired.*

Stasha, S. (2021). Amazon Alexa statistics, facts, and trends. *PolicyAdvice.net.* Last modified: July 25, 2023, https://policyadvice.net/insurance/insights/amazon-alexa -statistics

Spaulding, S. (2020). What is mindreading? *Interdisciplinary Review of Cognitive Science.* https://wires.onlinelibrary.wiley.com/doi/10.1002/wcs.1523

Stanovich, K.E. (2009). *What intelligence tests miss: The psychology of rational thought.* Yale University Press.

Stolcke, A., Ries, K., Coccaro, N., Shriberg, E., Bates, R., Jurafsky, D., Taylor, P., Martin, R., Meteer, M., & Van Ess-Dykema, C. (2000). Dialogue act modeling for automatic tagging and recognition of conversational speech. *Computational Linguistics, 26*(3), 339–371.

Strassel, S., Adams, D., Goldberg, H., Herr, J., Keesing, R., Oblinger, D., Simpson, H., Schrag, R., & Wright, J. (2010). The DARPA Machine Reading Program—Encouraging linguistic and reasoning research with a series of reading tasks. *Proceedings of the Seventh International Conference on Language Resources and Evaluation.* European Language Resources Association.

Sun, R. (1995). Robust reasoning: Integrating rule-based and similarity-based reasoning. *Artificial Intelligence, 75*(2), 241–296.

Sun, R. (Ed.). (2023). *The Cambridge handbook of computational cognitive sciences.* Cambridge University Press.

Sun, R., & Alexandre, F. (Eds.). (1997). *Connectionist symbolic integration: From unified to hybrid approaches.* Lawrence Erlbaum Associates.

Sun, R., & Bookman, L. (Eds.). (1994). *Computational architectures integrating neural and symbolic processes.* Kluwer Academic Publishers.

Sun, R., Slusarz, P., & Terry, C. (2005). The interaction of the explicit and the implicit in skill learning: A dual-process approach. *Psychological Review, 112*(1), 159–192.

Thayer, J. F., Hansen, A. L., Saus-Rose, E., & Johnsen, B. H. (2009). Heart rate variability, prefrontal neural function, and cognitive performance: The neurovisceral integration perspective on self-regulation, adaptation, and health. *Annals of Behavioral Medicine*, 37: 141–153. https://doi.org/10.1007/s12160-009-9101-z

Touvron, H., Lavril, T., Izacard, G., Martinet, X., Lachaux, M.A., Lacroix, T., Rozière, B., Goyal, N., Hambro, E., Azhar, F., & Rodriguez, A. (2023). Llama: Open and efficient foundation language models. *arXiv:2302.13971*.

Traum, D. R. (1999). Computational models of grounding in collaborative systems. *Working Notes of the AAAI Fall Symposium on Psychological Models of Communication*, pp. 124–131.

Traum, D. R. (2000). 20 questions for dialogue act taxonomies. *Journal of Semantics*, *17*(1), 7–30.

Turney, P. D., & Pantel, P. (2010). From frequency to meaning: Vector space models of semantics. *Journal of Artificial Intelligence Research*, *37*, 141–188.

Valmeekam, K., Olmo, A., Sreedharan, S., & Kambhampati, S. (2022). Large language models still can't plan (A benchmark for LLMs on planning and reasoning about change). *arXiv:2206.10498 [cs.CL]*.

Veale, T. (2012). *Exploding the creativity myth: The computational foundations of linguistic creativity*. Bloomsbury Academic.

Vrbanec, T., & Meštrović, A., (2020). Corpus-based paraphrase detection experiments and review. *Information*, *11*(5), 241. https://doi.org/10.3390/info11050241

Yuret, D. (1996, February 13). *The binding roots of symbolic AI: A brief review of the CYC project*. MIT Artificial Intelligence Laboratory.

Wilks, Y. (2019). Moral orthoses: A new approach to human and machine ethics. *AI Magazine*, *40*(1). Accessed June 6, 2023, link.gale.com/apps/doc/A581731439/AONE?u=nysl_oweb&sid=sitemap&xid=3352539e

Williams, T., & Scheutz, M. (2017). Referring expression generation under uncertainty: Algorithm and evaluation framework. *Proceedings of the 10th International Conference on Natural Language Generation*, pp. 75–84. Association for Computational Linguistics.

Winther, R. G. (2016). The structure of scientific theories. In E. N. Zalta (Ed.), *The Stanford encyclopedia of philosophy* (Winter 2016 ed.). https://plato.stanford.edu/archives/win2016/entries/structure-scientific-theories

Zanzotto, F. M. (2019). Human-in the loop artificial intelligence. *Journal of Artificial Intelligence Research*, *64*(1), 243–252.

Zhu, Z., & Hu, H. (2018). Robot learning from demonstration in robotic assembly: A survey. *Robotics*, *7*(2), 17. https://doi.org/10.3390/robotics7020017

Index

Note: Page numbers followed by *f* or *t* indicate figures or tables, respectively.